The Basics of INVESTIGATING FORENSIC SCIENCE

A Laboratory Manual

The Basics of INVESTIGATING FORENSIC SCIENCE

A Laboratory Manual

Kathy Mirakovits
Portage Northern High School
Portage, Michigan, USA

Gina Londino
Indiana University Purdue University
Indianapolis, Indiana, USA

CRC Press is an imprint of the
Taylor & Francis Group, an **informa** business

CRC Press
Taylor & Francis Group
6000 Broken Sound Parkway NW, Suite 300
Boca Raton, FL 33487-2742

Printed on acid-free paper
Version Date: 20150727

International Standard Book Number-13: 978-1-4822-2315-6 (Paperback)

Visit the Taylor & Francis Web site at
http://www.taylorandfrancis.com

and the CRC Press Web site at
http://www.crcpress.com

Dedication

I would like to dedicate this laboratory manual to Jay A. Siegel. You have been my mentor and my inspiration, constantly encouraging me to push myself to that higher level. I thank you for your constant support, but most of all for your friendship.

KATHY MIRAKOVITS

I would like to dedicate this laboratory manual to Jay A. Siegel, who has inspired me to believe that I can do anything and for the continual support and acknowledgment that I am a great educator.

GINA LONDINO

Contents

Unit 3 Forensic Biology

Preface

Forensic science education has expanded greatly in the academic world. The coursework used to be confined to four-year colleges and graduate schools, but currently forensic science classes can be found in local high schools as well as in two-year community colleges. This set of laboratory investigations is designed for the beginning forensic science student and for instructors who wish to have a solid laboratory foundation in basic forensic science topics.

The book is divided into five units. Unit 1—"Forensic Science and Crime Scene Investigation," which includes exercises in evidence collection, sketching a crime scene, and other basic tenants in crime scene investigation; Unit 2—"Patterns and Impressions," where students learn fingerprinting, toolmark and shoeprint analysis, firearm basics, and questioned document examination; Unit 3—"Forensic Biology," covers blood and bodily fluid analysis, pathology, anthropology, hair, and DNA; Unit 4—"Forensic Chemistry," investigates inks, fibers, polymers, drugs, arson, and explosives; and Unit 5—"Forensic Physics," studies glass, skid marks, and hair diameter. The laboratory investigations are not sequential and some are more advanced than others. The authors patterned the sequencing of the investigations on the forensic science textbook *Forensic Science: The Basics*, 3rd edition; however, the manual can be used either with the textbook or as a stand-alone unit. Laboratory exercises include enough background information for students to understand the topic being investigated; however, learning would be enhanced by using a text with the lab book.

The book includes extensive notes for instructors that will assist in pre-laboratory preparation. Most laboratory exercises have pre- and post-laboratory questions for students, basic crime scene scenarios, and lab objectives for instructors. Many of the exercises also have additional advanced lab exercises for educators who may have access to more specialized equipment.

The authors of this lab manual have a combined experience of more than 25 years of teaching laboratory forensic science at the basic level. To that end, the equipment needed to do most of the experiments is very basic and affordable for the undergraduate or public school educator. Suggested materials that can be ordered for use with some of the labs are referenced, including web addresses and product code numbers. Many of the laboratory investigations could be enhanced by using more elaborate equipment, if available.

We hope you enjoy using this manual to introduce the basics of forensic science to the next generation forensic science student.

Kathy Mirakovits
Portage Northern High School
Portage, Michigan

Gina Londino
Indiana University Purdue University
Indianapolis, Indiana

Authors

Kathy Mirakovits teaches forensic science and physics at Portage Northern High School in Portage, Michigan, and physics at Kalamazoo Valley Community College in Kalamazoo, Michigan. She holds a master's degree in science education from Western Michigan University and a bachelor's degree in science education from Miami University, Oxford, Ohio, and has completed graduate hours in forensic science. She has taught at the high school and at two-year-college levels for a total of 25 years and during that time has taught general science, physical science, chemistry, biology, earth science, and physics. Additionally, Kathy conducts workshops across the United States for teachers who wish to learn the application of forensic science in a school curriculum. She has developed numerous forensic science educational products for a national science supplier and has lead workshops at the National Science Teachers Association in forensic science.

Kathy served as president of the Michigan Chapter of the American Association of Physics Teachers and served as a curriculum writer for the Michigan Department of Education. Kathy currently serves as the high school director for the Michigan Science Teachers Association. She has received the RadioShack Science Teaching Award and is a state finalist for the Presidential Award for Excellence in Math and Science Teaching.

Gina Londino received her bachelor's degree in chemistry from Ball State University in Muncie, Indiana. She then attended graduate school at Purdue University, Indianapolis, where she earned a master's degree in chemistry. While in graduate school, her main focus was in forensic analysis of pigmented ink using pyrolysis gas chromatography-mass spectrometry. Gina also performed research on biomarkers at Eli Lilly and Company, Indianapolis, through the Adsorption Distribution Metabolism Extinction (ADME) department throughout graduate school. Gina is currently a senior lecturer in the Forensic and Investigative Sciences program at Indiana University–Purdue University, Indianapolis, where she has been teaching introductory courses in forensic science, forensic chemistry, and forensic microscopy since 2006. She has designed multiple courses in forensic science, including a variety of beginner-to-advanced laboratory exercises.

Unit 1

Forensic Science and Crime Scene Investigation

Lab 1

Introduction to Crime Scene Investigation

Solving the Puzzle

1.1 Introduction

Crime scene investigation is problem solving. Something has occurred that is out of the ordinary, and those given the task of investigating were not present to witness the event. A team of forensic scientists reconstruct what could have occurred based on evidence left behind at the scene. In order to do so, the team must work together as a cohesive unit, with each member contributing his or her individual talents and expertise to the investigation.

Effective investigations follow documented procedures that can be explained in court. In scientific investigations, the recommended procedure that is followed is termed as the *scientific protocol*. Protocols for collection of individual evidence from the scene are rules for proper handling of evidence, so that it is not compromised or contaminated. A crime scene team has a standard operating procedure or protocol established for how a scene is processed. When all team members follow the same set of guidelines or protocol, everyone understands his or her role in the investigation and the team works more efficiently and effectively.

In this laboratory activity, you will be presented with a *problem*—a puzzle that needs to be put together. As a team, you are to solve the problem of putting the puzzle back together, but more importantly, the team needs to think out *how* it goes about completing the task.

1.2 Pre-Laboratory Questions

1. What is a protocol?

2. Why is protocol necessary in problem solving?

3. Why is it beneficial to have more than one person investigating a crime scene?

4. Is following a procedure unique to crime scene investigation? Explain.

1.3 Scenario

Police recovered a picture that has been reduced to individually shaped pieces. Your team must restore the picture to its original state. As you complete the task, answer the questions listed in the procedure.

1.4 Materials

- Puzzle Box
- Pencil or Pen

1.5 Procedure

1. Before beginning the exercise, discuss methods that can be used to solve a puzzle. Design a plan that the team will use to complete the puzzle. Record the plan in Section 1.7, Part A.
2. As you put the puzzle together, discuss as a team how completing the puzzle is similar to investigating a crime scene. List these in Worksheet 1.7, Part B.
3. As you put the puzzle together, discuss as a team how completing the puzzle is different from investigating a crime scene. List these in Worksheet 1.7, Part C.
4. After completing the puzzle, review any problems or changes to the plan that had to be made. List these in Worksheet 1.7, Part D.

1.5.1 Venn Diagram

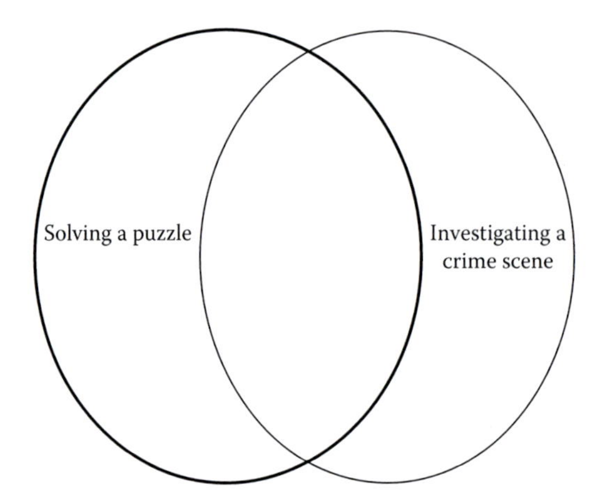

1.6 Follow-Up Questions

1. Class discussion: Share your team's plan, changes to your plan, and the similarities and differences in crime scene investigation with the class.

2. Complete the Venn diagram and answer the summary question.

3. In this exercise, solving a puzzle is compared to investigating a crime scene. Comment on the similarities and differences between what was done in the activity and the methods used to conduct *any* scientific investigation.

1.7 Worksheet

Discussion Points—Putting a Puzzle Together

Introduction to Forensic Investigation: Basic Problem-Solving Exercise
Team members:
Part A: Puzzle plan
Part B: Similarities
Part C: Differences
Part D: Review

Lab 2

Locard's Exchange Principle

Transfer of Evidence

> Wherever he steps, whatever he touches, whatever he leaves, even unconsciously, will serve as a silent witness against him. Not only his fingerprints or his footprints, but his hair, the fibers from his clothes, the glass he breaks, the tool mark he leaves, the paint he scratches, the blood or semen he deposits or collects. All of these and more bear mute witness against him. This is evidence that does not forget. It is not confused by the excitement of the moment. It is not absent because human witnesses are. It is factual evidence. Physical evidence cannot be wrong, it cannot perjure itself, it cannot be wholly absent. Only human failure to find it, study and understand it can diminish its value.
>
> Professor Edmond Locard
>
> **Paul L. Kirk**
> *Crime Investigation: Physical Evidence and the Police Laboratory, 1953*

2.1 Introduction

Edmond Locard, the famous French criminalist, once posited that every contact between person(s) and object(s) results in an exchange of evidence between those in contact. This became known as *Locard's exchange principle*, which is still a central principle of the analysis of scientific evidence that arises from civil and criminal activity. Although some of the exchanged evidence is overlooked because of its small size, much of it is recovered and falls into the category of scientific evidence known as *trace evidence*. Trace evidence can be classified in a number of ways and there is general agreement that it includes such items as hair, fibers, wood, glass, paint, soil, and explosive and fire residues.

Probably the most common type of trace evidence encountered in criminal activity today is textile fibers. Fibers occur in a huge variety of products, so it is not surprising that they occur in many types and instances of crime. Some estimates put fiber evidence in one-quarter of all crimes. At the same time, it must be noted that mass production of garments and other fiber-containing products means that it is ordinarily not possible to individualize a loose fiber to a particular source. Therefore, fibers, in most cases, are considered *class* and not *individual evidence*.

2.2 Pre-Laboratory Questions

1. How does Locard's exchange principle impact crime scene investigation?
2. What types of materials are easily transferred from one object to another?
3. Why are textile fibers common at a crime scene?
4. Can most fibers be classified as individual or class evidence? Why?

2.3 Scenario

Four individuals claim they were not in contact with each other. In fact, they have not seen each other for a week. It is your job to decide if their statement is factual or not, based on evidence.

2.4 Materials

- Woolen blanket (fire blanket works well)
- Translucent tape or fingerprint lifting tape
- Hand lens or microscope

2.5 Procedure

1. Within your group, take a *tape lift* from the back of each member's shirt between the shoulder blades. A tape lift involves placing a length of tape (~5 cm) on the back of the shirt and lifting the tape to remove any trace evidence.
2. Enter the value of the initial tape lift in the table for each group member.
3. One group member should take the woolen blanket and vigorously rub it on his or her back between the shoulder blades. After contacting the blanket, the same person should rub backs with another student. This second student then rubs backs with a third student, and so forth, until all members have rubbed backs.
4. Take another tape lift on each member's shirt as was done in step 1. Enter the value of this tape lift in the After Contact column in the table.
5. Compare the types of trace evidence found in the initial and after columns of the table.
6. Use a hand lens or microscope to compare and/or identify the evidence acquired.

2.6 Follow-Up Questions

1. What kinds of evidence were found in the initial tape lifts for each group member?

2. What kinds of evidence were found in the after contact tape lifts for each group member?

3. List any evidence that was *not* present in the initial lift and trace its path. How far did the evidence transfer?

4. List any evidence that *was* present in the initial lift that was transferred to other members. Show the path of transfer.

5. How does this activity illustrate Locard's exchange principle?

6. Tracy, a student in the classroom, has cat hair on her shirt. She states that she does not own a cat and neither do any of her friends. List the different ways that she could have gotten this hair on her shirt.

2.7 Locard's Exchange Principle Worksheet

Table of Evidence—Locard's Exchange Principle

Group Member	Initial Lift	After Contact	Items Identified
			Initial lift: After contact:
			Initial lift: After contact:
			Initial lift: After contact:
			Initial lift: After contact:

Lab 3

Physical Matches

3.1 Introduction

There are many types of evidence where physical matches, or fracture matches, can be determined. Different types of materials will exhibit unique characteristics when separated or fractured. From analysis of evidence such as this, conclusion as to whether or not multiple fragments may have once been whole or can be physically matched can be determined. The conclusion of the examination can be a match, be a nonmatch, or be inconclusive.

Before you make a physical match, there has to be a form of separation of a single item. Separation can result when an item is broken, torn, or cut apart into two or more pieces in a random fashion. An analyst can closely analyze fractured edges in order to make physical matches. The fact that breakages are random makes it possible to see unique and individual characteristics between two or more pieces. Because of these unique qualities along a breakage, a match can be determined or excluded; however, pieces that are too small to be physically possible to be put together can be inconclusive. In general, physical comparison is conducted between two questioned pieces of material between which specific physical characteristics are looked for and identified.

Surface features are simply any image, lettering, color, patterns, or markings that can be seen on the surface of a material. Whether or not the surface feature's details match up can help determine the likelihood of a match between two materials.

In forensic science, determining physical matches can be very important in the examination of evidence. Samples taken from crime scenes can be physically compared not only with each other but also with samples taken from possible suspects and other known samples. Just because two pieces do not have a match does not mean that the pieces could not have originated from the same source.

3.2 Pre-Laboratory Questions

1. What kind of material can produce a physical match?

2. Name one way an item can be separated that can result in a physical match.

3. Name a conclusion that you can reach when trying to determine physical matches.

4. Name an example of a surface feature.

3.3 Scenario

Jonathan and Jocelyn Jones, siblings who attend school together, threw a ping pong competition last weekend. Several classmates and neighbors attended the party and eventually got out of control. The police were called by fellow neighbors due to the noise of the party, and they observed several pieces of evidence upon their arrival. There were broken materials across the floor along with torn-up documents and fabric materials. A couple of the party attendees were taken into custody for disorderly conduct and evidence was also obtained to determine their identification.

3.4 Materials

- Playing cards cut up
- Fabrics cut up
- Wooden pencils broken

3.5 Procedure

1. Playing cards
   a. Match the pieces (1a) together based on their edge patterns or surface features and determine if the piece (1b) is a match.
   b. Record the color, shape, and number of pieces.
   c. Record any observations about the edges or patterns for how the pieces fit together.
   d. Sketch the final item to show the pieces as a whole.
2. Fabrics
   a. Compare the fabric pieces from the scene (2a) to the clothes of the people brought into custody (2b) and determine if they are from the people in custody.
   b. Record the color, pattern, and number of pieces.
   c. Sketch the final item to show the pieces as a whole.
3. Pencil
   a. Match the pieces of pencil together (3a) to form a complete pencil. Use the individual characteristics such as outer surface features and shape of the broken pieces to put the pencil together.
   b. Determine if the pencil was a physical match.
   c. Sketch the final item to show the pieces as a whole.

3.6 Follow-Up Questions

1. If a physical match can be made, what conclusion can be drawn?

2. Did all of your pieces in each section originate from the same object? Explain.

3. If you tore a blank piece of paper out of a notebook along the perforated edge, could you make a physical match from the notebook to the page?

3.7 Physical Matches Worksheet

Description of the Evidence	Sketch of the Evidence
Item 1a + 1b (cards):	*Item 1a + 1b:*
Were they a match?	
Item 2a + 2b (fabric):	*Item 2a + 2b:*
Were they a match?	
Item 3a + 3b (pencil):	*Item 3a + 3b:*
Were they a match?	

Lab 4

Evidence Collection

Making an Evidence (Drug) Fold

4.1 Introduction

Before processing of evidence can be done, the evidence at the crime scene must be collected and preserved. Determining what items at the crime scene are important and how those items should be collected is a skill practiced by crime scene investigators. They are trained to locate important items found at the crime scene, which are known as *evidence*. The duties of a crime scene investigator include not only being able to locate evidence but also how the evidence should be collected and preserved for transport to the forensic laboratory for analysis.

During the collection and preservation processes, it is important to understand how certain items should be collected and packaged. If an item is wet, it should not be placed in a plastic bag due to contamination by molding and bacterial growth. Additionally, if items are small, they should be placed in smaller containers. Fire debris evidence should be placed in air tight cans. Once the evidence has been properly contained, the storage container (whether a plastic bag, a paper envelope or bag, or a metal can) needs to be sealed with tamper proof evidence tape. This special tape will easily tear or deform when it is removed; therefore, it will be noticeable if anyone tampers with the sealed evidence.

The following exercise teaches how to make a simple container for small pieces of evidence. The container, called an *evidence fold* (or *druggist fold*), can be done easily in the field with a piece of paper or paper-like material. Paper is an ideal collection material for most items, as it allows the collected material to breathe, so molding is not encouraged. The evidence fold can be used to package and preserve evidence such as hair, fibers, small shards of glass, bullets and cartridge casings, and cigarette butts. It is important to remember, however, that each type of evidence should be placed in its own separate container. Do not mix items.

4.2 Pre-Laboratory Questions

1. When collecting evidence from a crime scene, what are the things to consider when deciding what type of packaging to use?

2. Why is evidence tape, or tamper proof tape, specifically used around the edges of a container?

3. Why is it not a good idea to place damp evidence in a plastic bag?

4. What types of evidence can be placed in an evidence fold?

5. Can more than one piece of evidence be placed in a container? If so, when can this be done?

4.3 Materials

- 8.5 × 11″ sheet of notebook paper or computer paper
- Pen

4.4 Procedure

1. Fold upper right edge of the 8.5 × 11″ sheet of paper across the paper until it meets the left edge. The result will be a triangle with an extra edge on the bottom that is untouched.

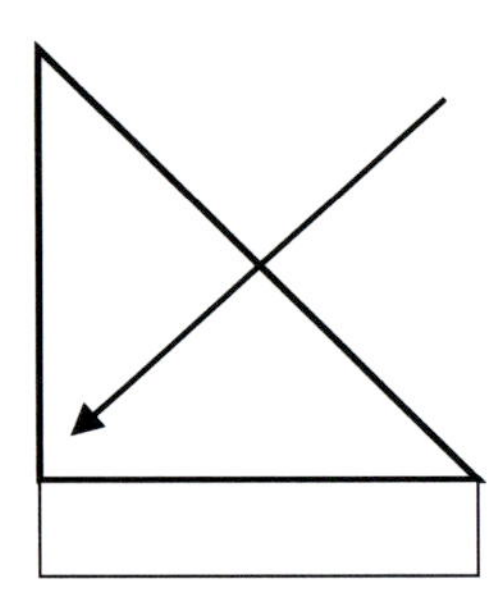

2. Fold the bottom edge up and make a good crease. Rip off the bottom edge, so that all that remains is the triangle and when the triangle is opened, the remaining paper will be a square.

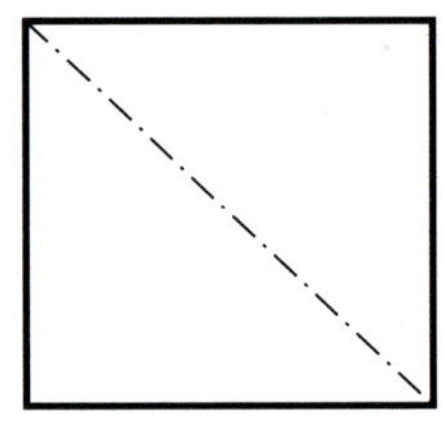

3. Refold the paper back into a triangle with the longer base at the bottom and the points at the top and the sides. Divide the base of the triangle into three equal lengths and fold the left and right sides of the triangle inward. The result should look like a box with a pointed top as shown below.

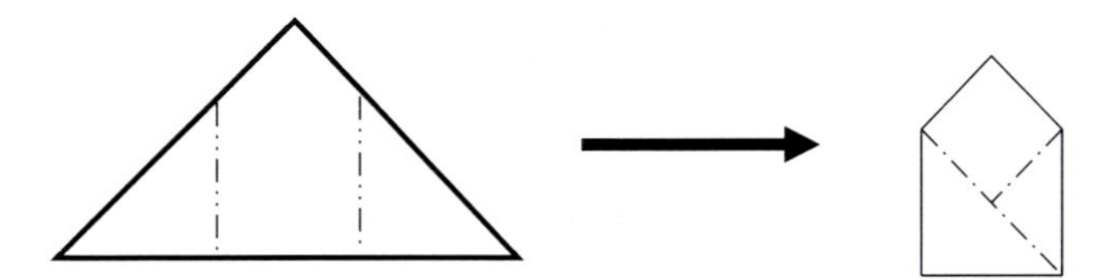

4. The pointed top of the evidence fold has two flaps that can now be opened to reveal a pouch for placing small pieces of evidence.
5. Take a small piece of the paper flap that was removed in step 2. This will serve as a pretend piece of evidence. Place the *evidence* in the pouch and push it all the way down to the base of the pouch.
6. The evidence is secured by making two folds and a tuck. First, fold the top point down, so that the remainder of the paper is now a square. Reopen this fold. Fold the square in half by folding *up* the bottom half where the evidence resides toward the pointed top.

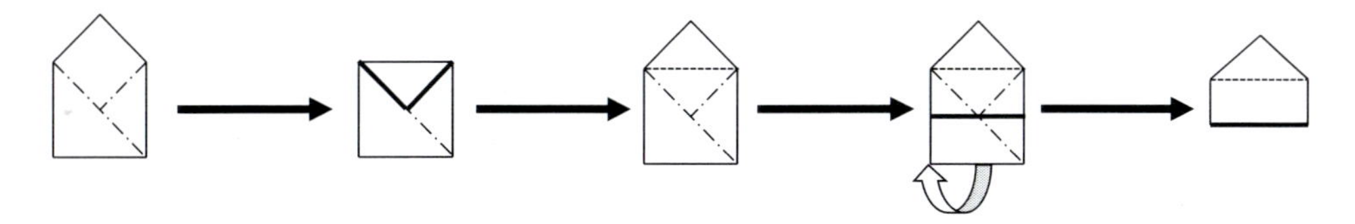

7. Open the top flap of the rectangle, and tuck the pointed top into the space where the two pointed end pieces cross. This secures the evidence in the fold.

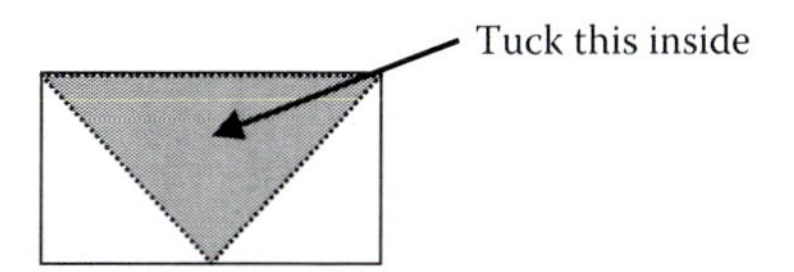

8. The fold would then be labeled with information such as who collected it, what type of evidence it is, where was it found, and the date/time it was collected. Next, it would be placed into a larger evidence envelope, labeled, and secured with tamper proof evidence tapes.

Lab 5

The 3Ls
Locating, Lifting, and Logging Evidence

5.1 Introduction

Did you ever wonder why crime scene investigators do not turn on room lights when searching for evidence? Did you ever wonder why they use flashlights when examining items from a crime scene?

The technique is called *oblique lighting*, and it is used to visualize trace evidence that would not be visible with regular overhead lights. Flashlights can direct lighting at any angle, specifically parallel to a surface. Light beams parallel to a surface make evidence that is loosely lying on top of a surface visible to the examiner. To demonstrate this, darken the room and shine a flashlight beam parallel across the classroom floor or the surface of a desk or lab table. Notice all the small particles, hairs, and fibers. Now turn the room lights on. Can you see the evidence with the overhead lights on?

In this activity, you will use a flashlight to locate trace evidence, collect it using forceps or tweezers, and place it in a druggist fold (see Unit 1: Lab 4). Evidence must then be placed in an envelope, labeled with important information, such as who collected the evidence, what *is* the evidence, where was it collected, and the date and time. The envelope will be sealed with tamper-resistant tape. After the evidence is collected, it is listed in a log sheet of evidence for transport to the crime lab. This log sheet establishes the chain of custody, a documentation of all the forensic analysts who handle the evidence.

5.2 Pre-Laboratory Questions

1. What kinds of evidence can be found using oblique lighting?

2. Why is it beneficial not to use overhead lights to illuminate an area and search for evidence?

3. What information should be placed on the outside of an evidence fold?

4. What is a chain of custody? Why is it so important?

5.3 Scenario

Police have collected items from a break-in at a local residence. Various items are in the crime lab and need to be processed for any trace evidence on the items. Each lab group is a team and must divide up duties as detailed in the procedure, locate evidence, package it properly, and log it into the evidence vault.

5.4 Materials

- Evidence to be processed for trace materials provided by instructor
- Flashlight
- Tweezers or forceps
- Paper for druggist folds
- Manila coin envelope or mailing envelope
- Manila envelope or paper lunch bag with *Evidence* label
- Marker or pen
- Tamper evident packaging tape or any sealing tape

5.5 Procedure

1. Each team should choose an *evidence officer*, who will log in the collected trace evidence. The remainder of the team will search for evidence, collect, and package it.
2. Lay the evidence to be searched flat onto a table.
3. Make a three to five druggists folds and have them at the table ready to package evidence.
4. Dim the lights and use the flashlight to search the evidence for trace materials. Begin the search in one corner and proceed systematically to each section of the material.
5. When trace evidence is located, use the tweezers to pick up the evidence and place it into the druggist fold.
6. Label the outside of the fold with the following information:
    a. Description of item
    b. Case number
    c. Date
    d. Location of collection
    e. Name of the collector

7. Place the labeled druggist fold into a manila coin envelope or mailing envelope. Label with the same information as above. Seal the envelope with tape and write your initials on the seal.
8. Place the evidence into an *evidence* package. Fill out the form, seal it, and log it in with the evidence officer.
9. The evidence officer should complete the *evidence log* for each piece of evidence, giving each piece of evidence an inventory number. That inventory number goes on the evidence label and the evidence log sheet.
10. Once all the evidence is collected and logged, the evidence officer takes the items with the log sheet to the crime lab director (your instructor) and fills out the chain of custody to turn over the evidence to the lab.

5.6 Follow-Up Questions

1. What are the possible outcomes if the chain of custody is broken or not complete?
2. What would you do if you located both hairs and fibers on one object in a crime scene?
3. What problems might arise if you have a wet or damp piece of evidence and place it into a plastic bag? How can you circumvent those problems?

5.7 The 3Ls Worksheet

Sample Evidence Label

Evidence

Case No. ___________ Inventory No. ____________

Type of offense______________________________

Description of evidence ______________________

____

Suspect ______________________________________

Victim _______________________________________

Date and time of recovery ____________________

____

Location of recovery _________________________

____

Recovered by _________________________________

Chain of Custody

Received from ________________________________

By ____

Date ________________Time ________________AM/PM

Received from ________________________________

By ____

Date ________________Time ________________AM/PM

Received from ________________________________

By ____

Date ________________Time ________________AM/PM

Received from ________________________________

By ____

Date ________________Time ________________AM/PM

Received from ________________________________

By ____

Date ________________Time ________________AM/PM

Received from ________________________________

By ____

Date ________________Time ________________AM/PM

Chain of Custody Log Sheet

Evidence Log Sheet

Law Enforcement Agency ______________________________

Evidence Officer ____________________ Case No. ____________________

Type of offense ______________ Date ______________ Time ______________ AM/PM

Suspect(s) ______________________________

Victim(s) ______________________________

Inventory No. ______________ Collected from ____________________

Inventory No. ______________ Collected from ____________________

Inventory No. ______________ Collected from ____________________

Inventory No. ______________ Collected from ____________________

Inventory No. ______________ Collected from ____________________

Inventory No. ______________ Collected from ____________________

Inventory No. ______________ Collected from ____________________

Inventory No. ______________ Collected from ____________________

Inventory No. ______________ Collected from ____________________

Inventory No. ______________ Collected from ____________________

Inventory No. ______________ Collected from ____________________

Inventory No. ______________ Collected from ____________________

Inventory No. ______________ Collected from ____________________

Total Number of Items ______________

Lab 6

Classifying Evidence

Is This Class or Individual Evidence?

6.1 Introduction

Prior to processing evidence, it must be sorted for efficient interpretation and application to the crime. The major classification schemes for evidence are as follows:

- Physical versus nonphysical
- Real versus demonstrative
- Known versus unknown
- Class versus individual

This activity will focus on deciding whether the evidence is class or individual.

6.2 Materials

- Unlined paper
- Pencil or pen
- Groups of items supplied by your instructor

6.3 Procedure

1. Look closely at one group of items. Classify the group by filling in the blank: "This is a group of ________." Decide if the group is class or individual.
2. Construct a flowchart on your paper of the evidence separating it by classifying it into categories until it cannot be divided into any more categories.
3. An example flowchart for birds follows. The bird group includes parakeet, crow, cardinal, and robin. Once the group is segregated, decide if the remaining item can be classified as individual or class. In the case of the robin, it is still class evidence.
4. Decide what would need to be analyzed for the item to be individual evidence. For example, an individual robin would be one that was tagged on the leg with an ID number, which is unique to that individual bird, or a robin that had a unique beak, leg markings, or other physical characteristic.

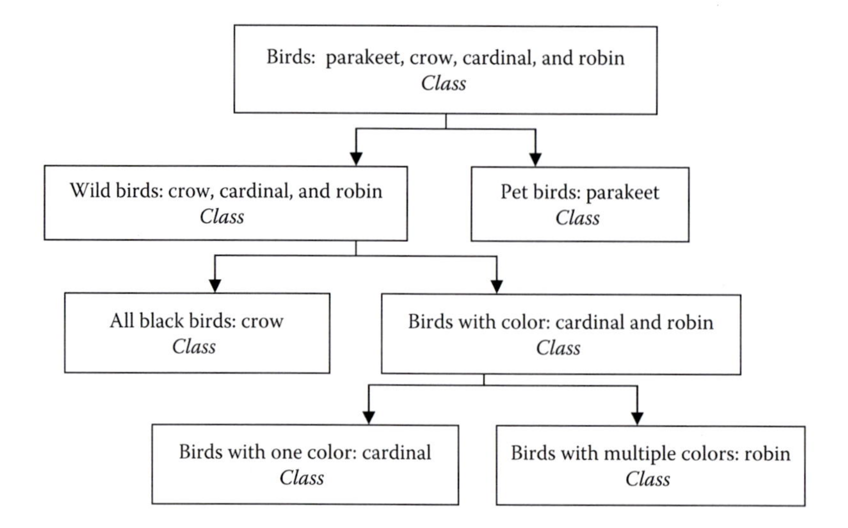

6.3.1 Flowchart

List the materials you are given: ______________________________.
Fill in the flowchart below to segregate your evidence. If you need to add or cross out boxes, do so.
Classify your evidence as *class* or *individual*. ______________________________.

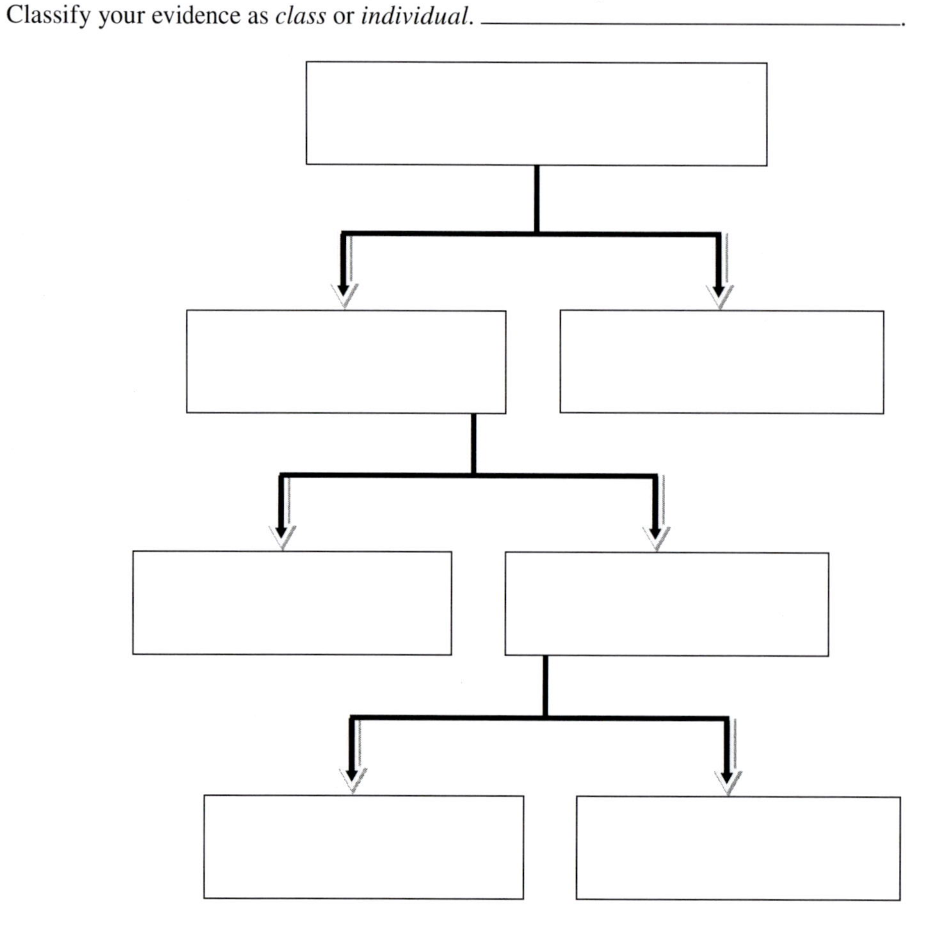

6.4 Follow-Up Questions

1. Blood is found at a crime scene and it is analyzed to identify the blood type. The blood type is determined to be of type AB+. Would this be class or individual evidence? Explain your answer.

2. Can the tread pattern for an unworn, size 10, pair of male athletic shoes be considered an individual characteristic? Why or why not?

3. List four examples each for individual evidence and class evidence.

Lab 7

Basic Crime Scene Sketch

7.1 Introduction

Quite frequently, investigators arrive to a crime scene after the crime has happened. In order to properly investigate the infraction, they must rely on evidence and what they observe at the scene. Before the evidence is lost or disturbed, the crime scene technicians must document how the scene *looked upon their arrival. This involves photography, videotaping, tagging evidence, more photography,* and then sketching the scene.

These steps are extremely important in order to get a picture of the scene, because it does not last forever. The final sketch should be an accurate picture of the scene that can be useful when questioning witnesses, suspects, and ultimately for court purposes. If the case goes to trial, the 12 jurors were not at the scene and in order to tell the story of what happened, the prosecutor needs to give the people of the jury the best possible picture of the crime scene.

A scale drawing is produced by taking linear measurements of the crime scene layout and evidence, so that they can be accurately placed in the scene. The first step is making a rough draft of the crime scene. The notes and measurements included in the draft will be the basis for the final crime scene sketch done off-site. The rough draft is drawn with a pencil on regular size paper and is not drawn to scale. The lines do not have to be perfectly straight, as all the sketching is done free hand. From the rough draft, the final crime scene sketch is completed usually on grid paper and must be drawn to scale using the measurements taken from the rough draft. It is important, therefore, to be as precise as possible in the rough draft measurements and to make the rough draft readable with all important measurements and evidence included.

The first step in the rough draft is to sketch the outline of the room, using most of the paper. Make it as large as possible. Then measure the room dimensions using feet and inches and put them on the sketch along with perpendicular end lines showing start and end points of the measurement. Feet and inches are used in crime scene sketches in the United States, since most jurors are more familiar with the English system measurements and not the metric system.

Sketch the location of doors and windows, including the width of each. Furniture in a scene should be roughly sketched to show relative location in the scene. Furniture can be a square, rectangle, or oval and does not have to be precise on its looks.

Evidence can be drawn as a circle with a letter or number in it that matches its designation in the list of evidence in the legend. Objects can be placed in a scene by taking two straight-line measurements from fixed, nonmovable objects, such as doors, windows, corners, trees, and telephone poles. When plotting evidence, various measuring methods can be used. One method uses rectangular coordinates, or straight lines, drawn parallel to the walls (see Figure 7.1a). A second method (Figure 7.1b) uses triangulation, with two lines to fixed points forming a triangle shape from the evidence. If evidence is strewn down a hallway, a baseline measurement system (Figure 7.1c) would be the most effective in placing the evidence. Objects are located a distance from a specific mark on

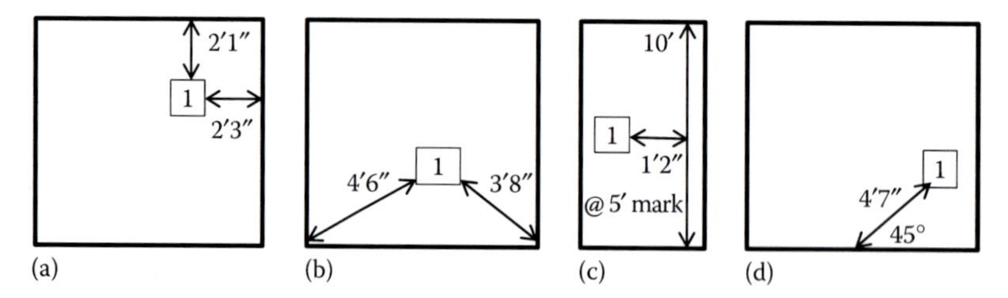

Figure 7.1
Locating evidence in a crime scene sketch: (a) rectangular, (b) triangulation, (c) baseline, and (d) compass points.

the baseline measure. Finally, evidence can be located with degree measurements from a compass or protractor (Figure 7.1d).

A legend or a list of evidence that includes the number or letter designated for the object and a brief description of what is placed along the perimeter of the paper. Along the top of the paper is a descriptive heading. The heading should include information on the upper left margin and information on the upper right margin. The upper left margin should include case number, address of the crime, date and time of the sketch, and the type of crime being investigated. The upper right margin should include the names of the important persons such as the names of the victim, the investigator doing the sketch, and the person who assisted in the sketch.

A compass direction must be added to your rough draft to show orientation. It is best practice to always have north directed toward the top of your paper, so you must orient yourself in the crime scene such that north is at the top, east to the right, west to the left, and south at the bottom.

Once the rough draft is complete, a final sketch drawn to scale is made on grid paper with a ruler. The scale for the drawing must be included, usually at the bottom of the page. Examples of a rough draft and a crime scene sketch are shown in Figures 7.2 and 7.3, respectively.

7.2 Pre-Laboratory Questions

1. Why is it important to sketch a crime scene?

2. How does one determine what type of measurement method to use?

3. What information should be included in the heading of the rough draft?

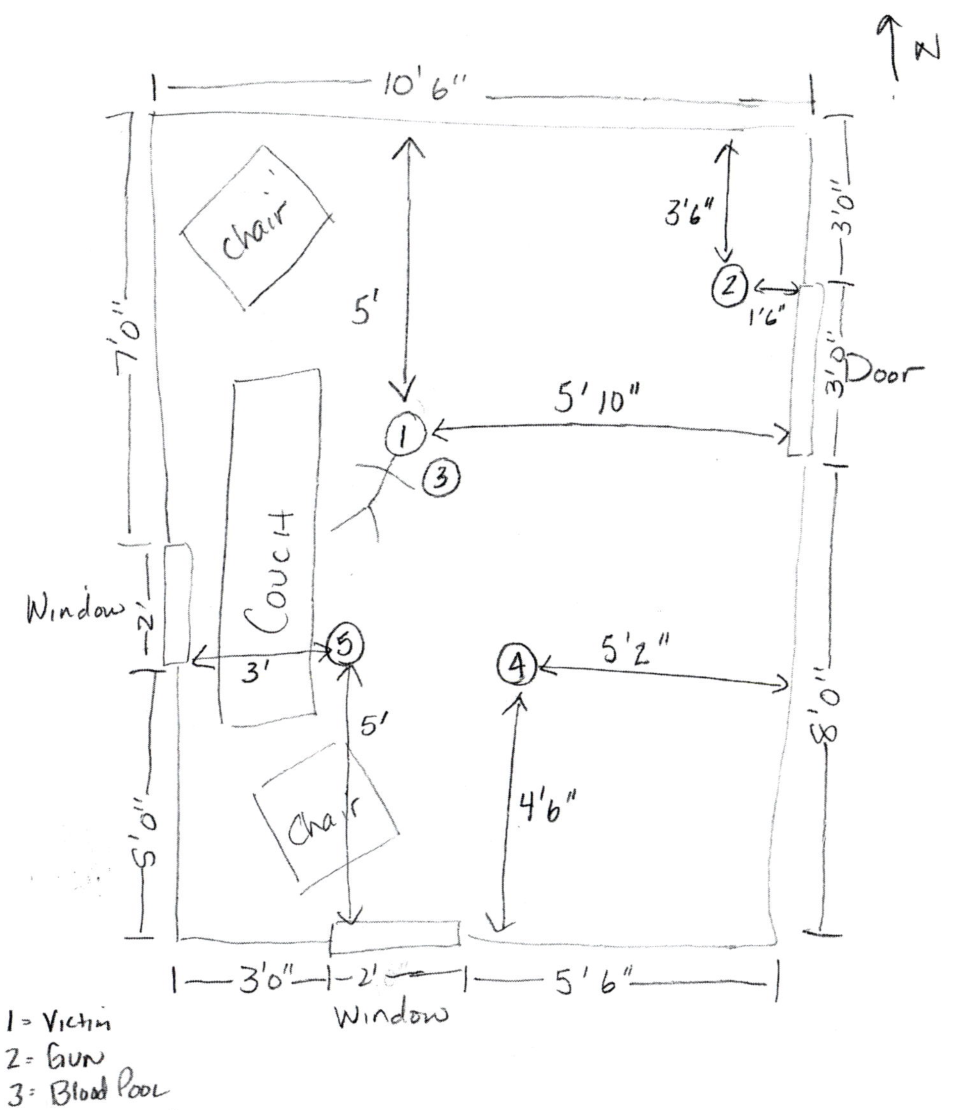

Figure 7.2
Rough sketch of the crime scene.

Case # 12345
1098 Main Street, Newtown, MI
05/14/2014; 8:49 PM
Homicide

Victim: John Dunn
Investigator: Mary Smith
Assisted by: Tom Bush

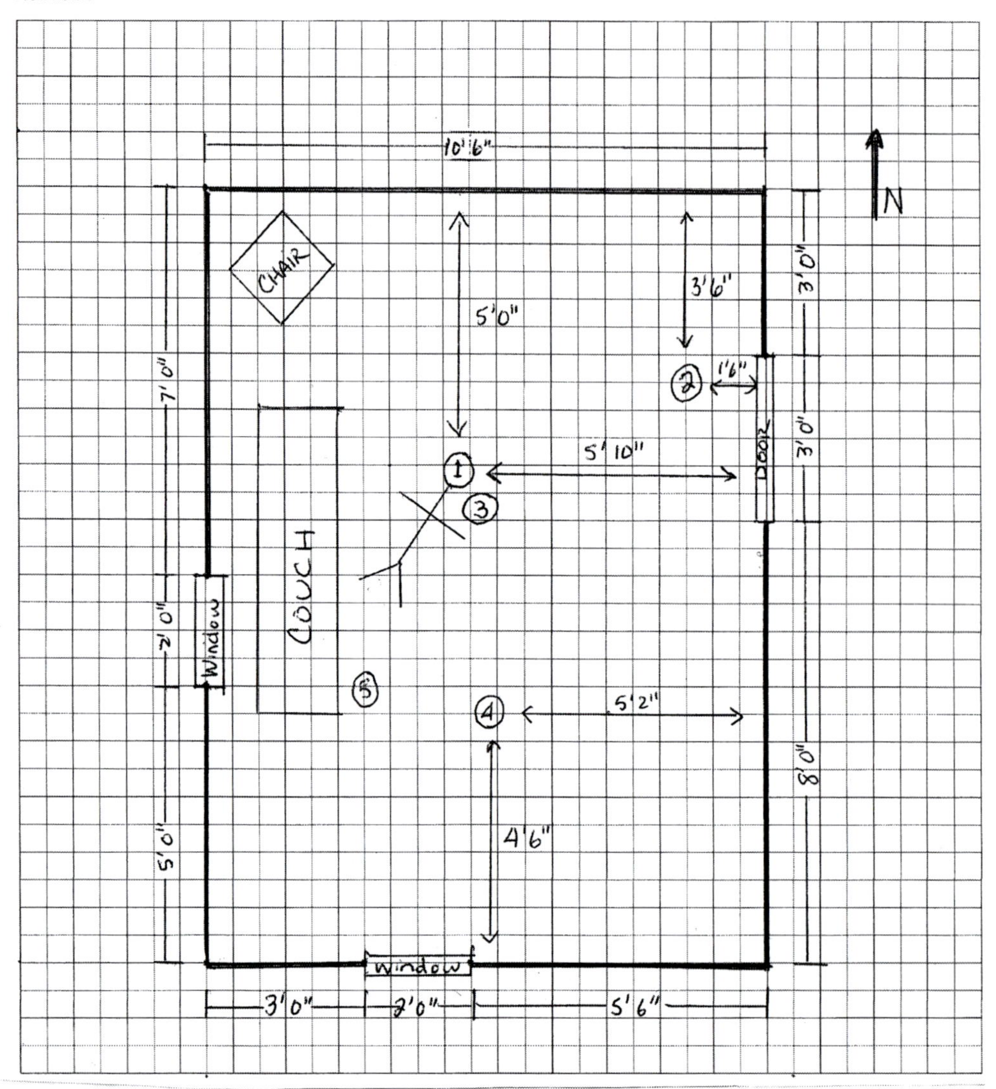

Legend
1—Victim
2—Revolver
3—Blood Pool
4—Cartridge Casing
5—Cigarette Butt

Figure 7.3
Final sketch of a crime scene in a grid paper.

7.3 Scenario

Your instructor will have a scenario and crime scene for you to sketch.

7.4 Materials

- Tape measure, meter sticks, and rulers
- Plain paper and graph paper (grid paper)
- Pencils and black ink pens

7.5 Procedure

1. *Rough draft*: See Figure 7.2, for example.
   a. Make a heading at the top of your paper. Include all the information listed in the introduction.
   b. Draw a rough outline of the shape of the room. This outline should take up most of the paper, leaving room at the bottom or side for a legend.
   c. Position north at the top of your paper and put in a compass showing north.
   d. Take perimeter measurements of the room and mark them in your sketch, using perpendicular line segments to show the start and end points of each measurement.
   e. Measure door and window widths and add them to the sketch.
   f. Tag important evidence if it has not already been done and make a list of the evidence for the legend.
   g. Use one of the evidence locating methods to measure and record the position of all the pieces of evidence.
2. *Final sketch*: See Figure 7.3, for example.
   a. Taking the measurements from the rough draft, determine a scale for your final sketch. For example, 1/4 inch on the sketch could be equivalent to 1 foot at the scene, or 4 squares on your grid could equal 1 foot in the scene.
   b. Use the scale and rulers to make a neat, precise final sketch. The final sketch should be in ink, not pencil, so that it cannot be altered and is more visible on the grid lines.
   c. If there is a great deal of evidence to be placed in the scene, not all pieces need to have measurements shown, since it is drawn to scale. Usually the major pieces of evidence include distance measures in the sketch, but others do not, to make the sketch less busy.
   d. Check your final sketch to be sure that it includes the following:
      i. Headings
      ii. Compass direction for N (north)
      iii. Legend
      iv. Scale
      v. Important measurements in feet and inches

7.6 Follow-Up Questions

1. Why would it be important to sketch a crime scene to scale?

2. Why is it necessary to know who sketched the crime scene and who assisted?

3. Would the date and time in the heading of the sketch be the date and time of the actual crime? Explain your answer.

4. Not all evidence has the location measurements in the final sketch. Why?

7.7 Basic Crime Scene Sketch Worksheet

Final Sketch Grid Paper

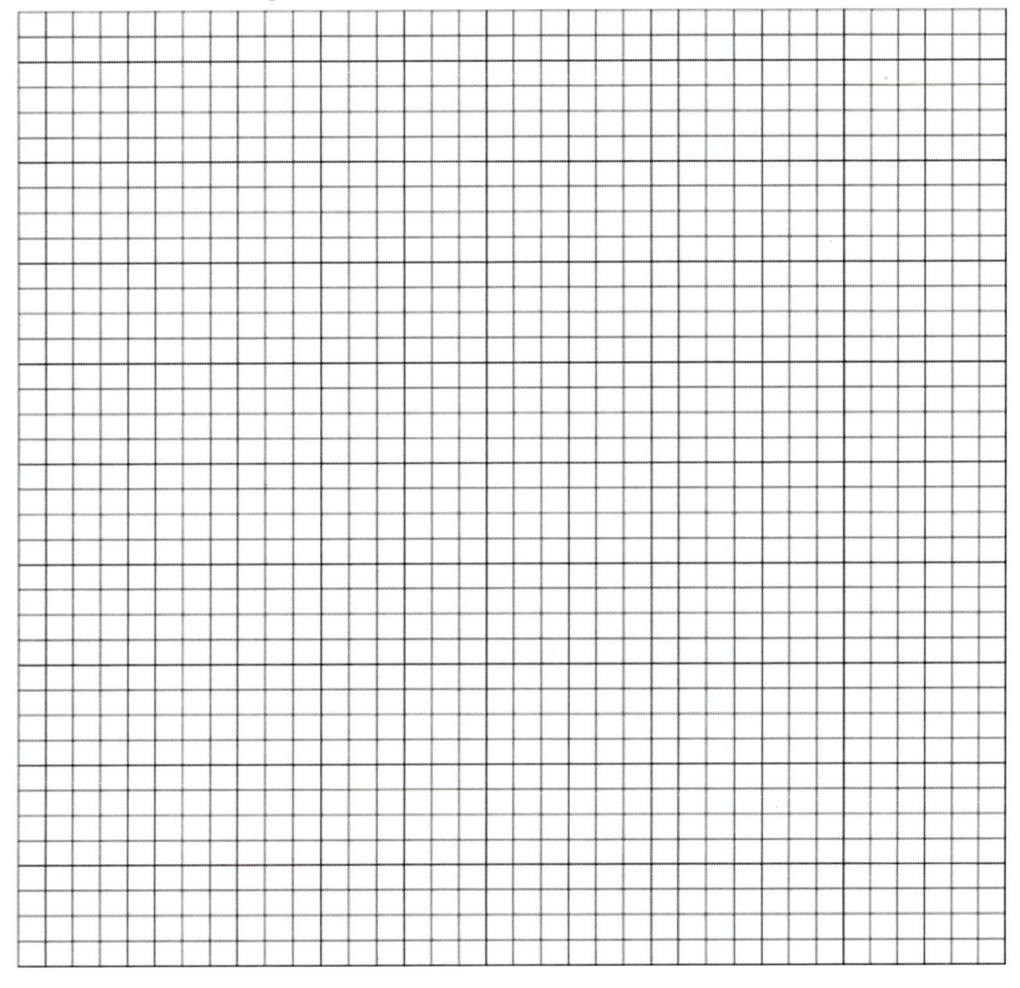

Lab 8

Crime Scene Investigation

8.1 Introduction

Before processing of evidence can be done, the evidence at the crime scene must be collected and preserved. Being able to determine what items at the crime scene are important and how those items should be collected is a skill. Crime scene investigators are trained to locate important items found at the crime scene known as *evidence*. The duties of a crime scene investigator include not only being able to locate evidence but also how the evidence should be collected and preserved so that the items may be transported to the forensic laboratory for analysis.

The first step is to be able to document the crime scene. This includes constructing a drawing of the scene to scale and recording either with photographs and/or video, the crime scene. Specific measurements need to be included of where items are found in relation to doorways, windows, a victim, and so on. Once evidence is located, photographs of the evidence should be taken in relation to other items within the crime scene, as well as photographs of the evidence up close with and without a scale. The purpose of documentation is to be able to re-create the crime scene at the forensic laboratory. The location and position of the evidence may be useful information to the forensic analyst during examination.

There are many considerations during a crime scene investigation. Determination of what is important evidence to the crime and what is not beneficial must be assessed. The type of method that should be used when collecting and preserving evidence must also be determined.

During this laboratory exercise, you will need to practice the power of observation and learn how to document and collect a variety of forms of common evidence found at crime scenes.

8.2 Pre-Laboratory Questions

1. What is the duty of a crime scene investigator?

2. What should be included in photographs taken of the evidence?

3. Define the chain of custody.

8.3 Scenario

Police respond to a call that reported a suspicious scene at the neighbor's residence. Officer Ray Lee and his partner arrived at the house and found the front glass door broken on the floor. Upon entry into the house, they noticed a broken window with a red substance on it. They found the owner, Joey Smith, with face down in the kitchen with a pool of brownish red liquid surrounding his head. Next to him was a pipe with blood and hair on it.

You, the crime scene investigator, have been brought to the scene. Your job is to properly collect, store, and document the evidence you find at the crime scene. You will need to determine the point of entry; identify the murder weapon; and collect evidence, which includes dusting for fingerprints, collecting any possible blood samples for DNA analysis, collecting any trace materials, and preserving anything else that can be used as evidence for future analysis.

The following documents will provide you with a way to record what you see. Be sure to properly document each evidence container with a chain of custody document. Take pictures of each piece of evidence with a scale and document your images in the photo log. Also, be sure to provide an accurate sketch of the crime scene in the crime scene grid. The crime scene evidence worksheet must be filled out for each piece of evidence collected. Ensure your descriptions are detailed and sketches draw attention to crucial aspects to each of the following item:

8.4 Materials

- Mini crime scene
- Camera
- Ruler

8.5 Procedure

1. Sketch the crime scene
   a. Using the grid paper provided, sketch the crime scene and label the distance between each of the three pieces of evidence and the point of origin; determine and give a key of number of squares per foot on the grid paper as personally determined.
   b. Start the sketch using the door as the point of origin and measure the distance between the door and a piece of evidence and draw or name that piece of evidence on the grid paper.
   c. Then measure and record the distances between each piece of evidence.
2. Photograph the evidence
   a. Take three photographs of a piece of evidence.
   b. Fill out the photo log for each picture stating the number of the picture, what it is, and the time and date.
   c. The first picture should be at a distance to show the orientation of the evidence to the room.
   d. The second picture should be an up close picture of the object with approximately 1 cm of a border around the object. Make sure to be directly over the evidence.
   e. The third picture should be taken with a ruler to show the scale of the object. Again, take the picture directly over the evidence.
   f. Repeat steps b through e for each piece of evidence collected.

3. Collect the evidence
   a. Fill out the chain of custody form for each piece of evidence collected. Each piece of evidence should have the date, item number, and initials of the collector on it on the container. Put on gloves before collecting the evidence.
   b. Collect the trace material from the pipe using tweezers and place in a plastic bag.
   c. Collect the blood from the pipe using a wet cotton swab and place in a microcentrifuge tube by breaking off the end of the swab in the tube.
   d. Collect the fingerprint from the pipe by lightly dusting the pipe with fingerprint powder. Then roll the tape onto the print making sure there are no air bubbles in the tape over the print and then gently lift the tape off of the print and place on area on paper.
   e. Collect the glass piece and place in an envelope and label.

8.6 Follow-Up Questions

1. Why is taking pictures with and without a scale important?

2. Why do you need to sketch the crime scene?

3. You collected a piece of evidence, but did not file the chain of custody document properly. What are the possible consequences?

4. Would you place pieces of broken glass in a plastic bag? Why or why not?

5. What additional equipment do you use when collecting evidence and why is it necessary to use them?

Unit 2

Patterns and Impressions

Lab 9

Fingerprints

9.1 Introduction

Fingerprints have both class characteristics and individual characteristics discoverable by the forensic scientist. The most commonly observed class characteristic of fingerprints is the general flow of the friction ridge pattern, which can consist of a loop, whorl, or an arch. A loop consists of a ridge pattern that enters from one side of the finger and curves around to exit out the same side of the finger. Loops always contain one *delta*, a divergence in the ridge flow, in the friction ridge pattern. A whorl contains at least one ridge that makes a complete elliptical pattern in the middle of the ridge pattern. Whorls contain two deltas in the friction ridge pattern. Arches have a ridge pattern that enters on one side of the finger and exit out from the other, and therefore, no deltas are present in the pattern (Figure 9.1).

Individual characteristics of fingerprints include the location of smaller features called *minutiae*. Enclosures, cores, deltas, bifurcations, islands, and ridge endings are all recognizable features that show up in different places in each individual fingerprint. Because the theory behind fingerprinting claims that no two fingerprints are alike, the locations of these recognizable features are characteristics that are individual to the specific fingerprint's friction ridge formation (Figure 9.2).

There are multiple ways to collect and preserve fingerprints depending on the circumstances of how the print is made. The prints can be latent, patent, or plastic. Latent prints are invisible prints on a surface that have to be recovered with powder. Patent prints are prints that are visible on a surface due to some sort of previous staining, such as from ink, blood, or grease. Plastic prints are prints that are formed as a 3D impression, such as an impression in clay. Due to the different ways prints can be seen, different techniques need to be utilized to properly collect and preserve them.

This set of experiments will give you practice in the following:

- Developing fingerprints from different surfaces using powders
- Developing fingerprints using cyanoacrylate (superglue) fuming
- Depositing your own fingerprints onto fingerprint cards

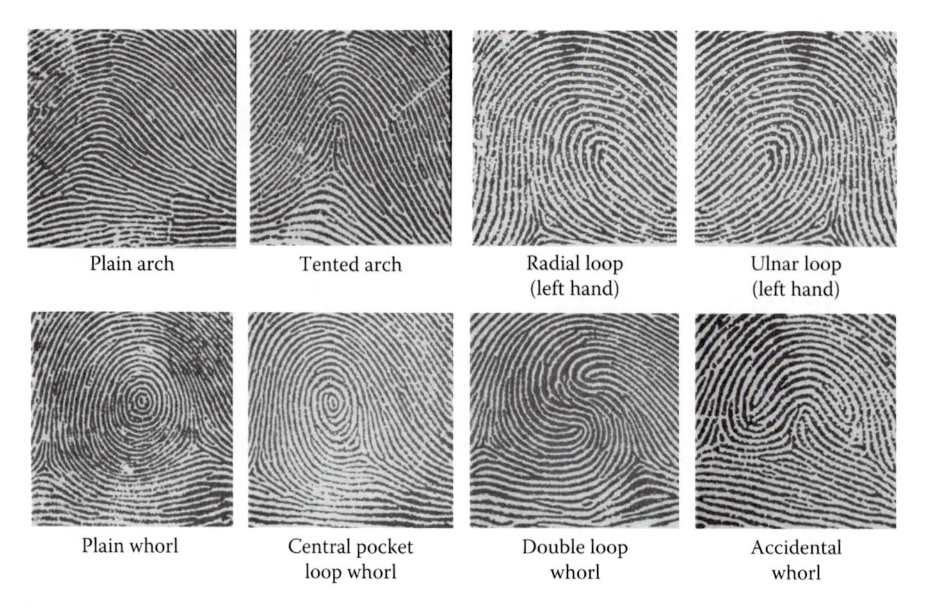

Figure 9.1
Class characteristics of fingerprints.

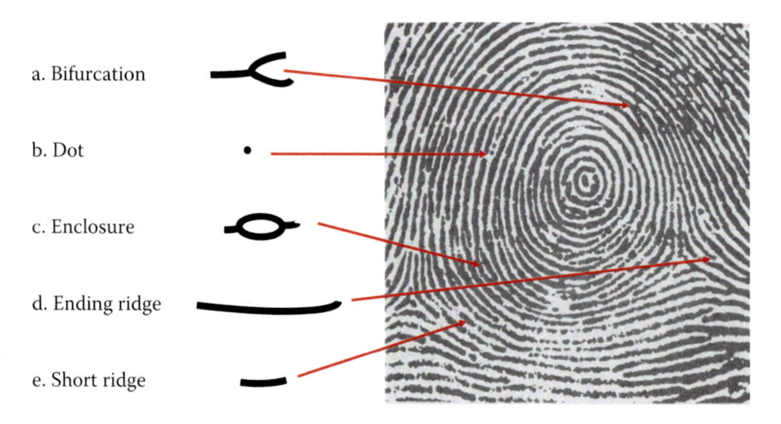

Figure 9.2
Individual characteristics of fingerprints.

9.2 Pre-Laboratory Questions

1. Name the three friction ridge patterns.

2. What is the difference between latent and patent fingerprints?

3. What type of fingerprint developing method is permanent?

9.3 Scenario

The art room at school recently displayed a new exhibit that featured several framed photographs from a famous photographer. The director, Brooklyn Petals, reported that a couple of pictures had been stolen from the exhibit and only the photographer had recently handled the pictures. The pictures were found in a box outside of the building and brought in as evidence. Students who had access to the building the previous night were fingerprinted and the prints were taken into the lab to be compared. Compare the prints collected from the pictures and compared them to the students' fingerprints.

9.4 Materials

- Black fingerprinting powder
- Fingerprint brushes
- Microscope glass slides
- Superglue fuming materials
- Fingerprint cards
- Ink pads

9.5 Procedure

1. Dusting for fingerprints
   a. Deposit a few fingerprints on a surface.
   b. Using black powder and a brush, lightly dust the fingerprint with the powder.
   c. Once the black powder has collected onto to the fingerprint, remove all the access powder.
   d. Take a piece of clear tape and place it over the dusted fingerprint.
   e. Remove the tape and place it on a white index card.
   f. Identify the pattern of the fingerprint.

2. Superglue fuming
   a. Deposit a few fingerprints on a microscope slide. Place hot plate into the chamber.
   b. Deposit about a nickel size of glue onto a small aluminum dish and place onto the hot plate. Add a small plastic cup of water to the chamber. Place the glass microscope slide inside the chamber, plug in the hot plate, and secure the lid on the chamber.
   c. Wait for 15 minutes, while observing every five minutes.
   d. Optional: After the evidence has been fumed, apply some of the Rhodamine 6G dye solution to the print with a swab to help visualize the print better and look at under ultraviolet light.
   e. Identify the pattern of the fingerprints.
3. Inked fingerprints
   a. Complete the fingerprint card on your own fingerprints using an ink pad.
   b. Exchange your fingerprint set and your lifted print on the index card with your lab partner.
   c. Determine which finger made the lifted print on your partner's fingerprint set.

9.6 Follow-Up Questions

1. What are advantages and disadvantages of using the black powder to develop latent fingerprints?

2. What features did you use to compare and match the lifted fingerprint?

3. When documenting a fingerprint for a legal record, what should you consider when making the fingerprint?

4. What is the purpose of developing fingerprints using the superglue fuming method?

9.7 Fingerprints Worksheet

Henry Classification					
Finger	Right Thumb	Right Index	Right Middle	Right Ring	Right Little
Ridge pattern					
Henry classification value					
Finger	Left Thumb	Left Index	Left Middle	Left Ring	Left Little
Ridge pattern					
Henry classification value					
Sum of numerator of Henry classification values (white boxes)					
Sum of denominator of Henry classification values (gray boxes)					
Henry classification of person					

Lab 10

Questioned Documents

10.1 Introduction

Document analysis includes examination of handwriting on a document that is suspected of being tampered with or forged. This consists of any object that contains linguistic or numerical markings in which there is doubt as to who the author is or if the writing is authentic.

Handwriting analysis is conducted on questioned documents through the theory that each person possesses unique characteristics when writing and therefore a person's writing cannot be exactly repeated by any other person.

Handwriting analysis can be conducted from two different types of writing exemplars: requested and nonrequested. Requested writings consist of writing samples that are taken with the intention of comparison to a questioned document. Since these writings are acquired upon request, the conditions by which these samples are obtained can be controlled. In addition to requested writings, nonrequested exemplars, or writings acquired from the normal course of business or personal transactions, can also be used to compare to questioned documents. The advantage of nonrequested writings includes that the provided writings cannot be attempted to be forged or altered when provided, since they are written without the knowledge of being used in comparison to a questioned document and therefore are more likely to truly reflect a person's handwriting. However, variables such as the writing instrument used, surface of writing, and text provided cannot be controlled unlike requested documents.

When analyzing provided exemplars of writing to a questioned document, class and individual characteristics present in the individual's writing are analyzed to determine if the known sample of writing could have come from the same person as the questioned writing. Characteristics such as spacing between letters and words, relative proportions between letters and within letters, individual letter formations, formations of letter combinations, overall slant of writing, connecting strokes, pen lifts, beginning and ending strokes, unusual flourishes, and pen pressure are examined between the known and unknown sample to analyze the similarity between the two samples. Overall, the purpose of handwriting analysis is to determine if an individual could have written a specific questioned document.

One tool used in questioned document examination is alternating lighting on the sample or a video spectral comparator (VSC). Specific wavelengths of light are introduced onto the document. Because different pens are made of different inks, different pens will reflect and absorb different wavelengths of light. Using this concept allows a forensic analyst to see when different inks were used on a document and determine how the document was altered.

10.2 Pre-Laboratory Questions

1. Name a characteristic that is used to compare handwriting samples.

2. A journal entry was used to compare the handwriting on a suicide note. What type of exemplar was the journal entry?

3. What can be used to look for alterations and obliterations on documents?

10.3 Scenario

Fredrick Harrison is the owner of an art museum downtown. He noticed that the funds from the museum were randomly disappearing from the account to institutes that he had no known connection. He felt like this was very suspicious and decided to have an investigation of the funds and records for the money transactions between the institutes. A document examiner asked for all of the written records that the museum had in their possession for the past year. After a visual examination of the checks, the examiner concluded that the signature on the some of the checks did not match with that of Mr. Harrison's. Also, on the checks that Mr. Harrison did sign, with the use of an alternate light source or a VSC, it was concluded that there were alterations made later to the amount that the museum was giving.

10.4 Materials

- Unknown questioned document
- Known author documents

10.5 Laboratory Procedures

10.5.1 Basic Laboratory Procedure

1. Handwriting comparison
   a. Take the provided questioned document and determine if the same person wrote the entire document. Compare both the victim's and suspect's handwriting exemplars to the document to help determine this.
   b. Look at the style of the writing, the shape of the letters, slope of writing, spacing between letters and words, initial, connection, terminal strokes, and so on.
   c. Determine and record why the document could or could not have come from the same person along with who wrote the document.

10.5.2 Advanced Laboratory Procedure

1. VSC or alternate light source
   a. Place the document under the light of the VSC.
   b. Examine the document for variations by noting any difference in the shades of color on the ink under different wavelengths of light.

c. Repeat steps a and b looking at the document with the three different light source wavelengths.
d. Determine if the document appears to express alterations and record what these changes are, if present, on the worksheet.

10.6 Follow-Up Questions

1. Looking back at your questioned documents, how could you tell which ink was written first and what words/numbers were originally written? Explain using a specific example from your experiment.

2. In your examination of the handwriting samples, what did you determine based on the comparison and why did they come from the same person or not?

10.7 Questioned Documents Worksheet

Describe both handwriting style samples. Are they from the same person? Why?
Describe what you saw when looking at your samples with the VSC. Include details such as wavelengths, what was added to the documents, what the document originally stated, and so on.

Lab

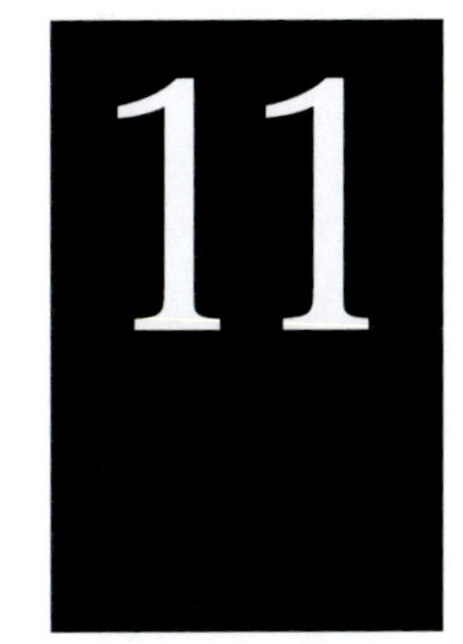

Firearms Identification Web Quest

11.1 Procedure

1. Go to the website http://www.firearmsid.com/.
2. Answer the following questions as you read the material on the site.
3. On the left side of the home page, find the MAIN MENU.
4. Click on Firearms Identification. Choose Introduction.
5. Answer the following questions as your make your way through the website.

11.2 Questions to Answer

11.2.1 Introduction to Firearms ID

1. What is the definition of firearms identification?

2. Firearms examiners test firearms to see if they work properly, and for evidence recovered, they determine the caliber and ________________________________ of bullets and cartridge casings.

3. Additionally, the examiner will receive clothing from detectives. Why?

11.2.2 Fundamentals of Firearms ID

1. What causes a particular firearm to be unique?

2. What percentage of firearms produces *mechanical fingerprints* that can be matched in forensic investigation?

3. List and briefly describe four class characteristics that are part of the matching process for firearms.

4. After the class characteristics match is verified, what individual characteristics are evaluated?

11.2.3 Bullet Identification

1. A bullet is evaluated by the rifling characteristics—land and grooves to match *class* characteristics. What does the examiner look for to determine an *individual* match for a bullet? Where are these found on the bullet?

2. Name two types of equipment used by the firearms examiner when testing a firearm for matching characteristics.

3. What is the one problem that the examiner might have with submitted evidence?

4. When reporting out results of the examined evidence, what are three types of reports that can be submitted?

   a.

   b.

   c.

   Scroll down to the bottom of the page to the Results section, and click on the "general rifling characteristics (GRC)" link.

5. These data tables are available to examiners to help match firearms. If the examiner's bullet had a right twist with six lands and grooves and a land width of 0.054 and a groove width of 0.126, which manufacturer did the bullet come from?

6. The Llama, Mauser, and Star bullets all have the same twist, number of lands and grooves, and land width. How are they different? Would precision in measurement be important for the examination? Why or why not?

7. Why are examiners interested in the manufacturer of the gun that fired the bullet? How can it aid an investigation?

11.2.4 Cartridge Case Identification

1. From what material are most cartridge casing made?

2. Name additional materials used in making cartridge casings.

3. Name and describe the two types of marks that can be found on cartridge casings.

4. How are the results of cartridge casing investigation reported?

5. Chamber marks are striations that occur on the side of the casing and most of them occur after the cartridge was fired. What process(s) produce the striations after shooting the weapon?

6. Look at the photo of the extractor. What is its function? Where will the extractor mark be found? Sketch the location below.

7. Name three types of impressed action marks.

8. What part of the cartridge has *breech* marks?

9. What action caused these breech marks?

10. Sketch and label one set of breech marks. Denote the location of the marks with an arrow.

11. What causes *chamber marks*? Sketch a cartridge casing and label the marks.

12. List at least four other marks that are used to identify a cartridge casing.

11.3 Case File Exercise

1. Go to the MAIN MENU.
2. Click on Firearms Identification.
3. Choose Case Profiles.
4. Choose Bullet and Cartridge Case Comparison.
5. Read the case and discuss how forensic firearms identification was crucial to the case.

Lab 12

Basic Firearms Identification
Examining Bullets and Cartridge Casings

12.1 Introduction

Firearms identification involves the analysis of toolmarks. When a firearm is manufactured, metal tools are used to carve out the barrel of the gun and to impart a twisting pattern called *rifling*, which gives the bullet spin. Whenever metal scrapes metal, it causes unique scratches and wear patterns on both instruments. Additionally, each time the metal tool is used, it changes, acquiring more scratches, nicks, and wear. This means the same tool can be used on many gun barrels; because each barrel is made up of metal, it imparts wear and scratches to the tool, making each barrel slightly different or unique.

These gun barrel toolmarks are then transferred to the surface of a bullet as it is fired through the barrel. Other markings are left on cartridge cases as a bullet is fired. These markings were originally made by metal tools that made the parts of the weapon.

12.1.1 Bullet Identification

When a weapon is fired at a target, the shooter would like the bullet to have the best chance to hit where it is being aimed. This means that the bullet must be made to spin on its long axis as it emerges from the barrel of the weapon. This is accomplished by manufacturing the barrel of the weapon so that rifling is incorporated. The rifling process bores the inside of a gun barrel from one end to the other, producing a series of lands and grooves (see Figure 12.1).

The *number* and *width* of lands and grooves, the *direction* of their twist through the barrel are all *class characteristics* that can give valuable information to the firearms examiner about the manufacturer and model of the weapon. The twist of the lands and grooves is noted as a right twist (clockwise) or left twist (counterclockwise). Additionally, the examiner determines the *caliber* of the bullet and the *type* of bullet. Figure 12.2 shows some of these class characteristics.

Caliber is the diameter of the barrel of the gun. Since the bullet closely matches the diameter of the barrel, the caliber of the bullet is the diameter of its base. The caliber of the cartridge case is

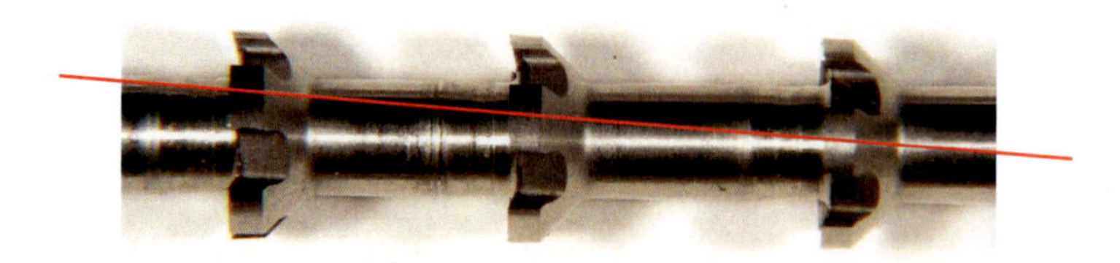

Figure 12.1
A rifling broach for gun barrel. Note the downward path to the notches that carve the grooves in the barrel. This imparts the twist in the bullet. (Courtesy of Chris Monturo, Precision Forensic Testing, Dayton, OH, www.precisionforensictesting.com.)

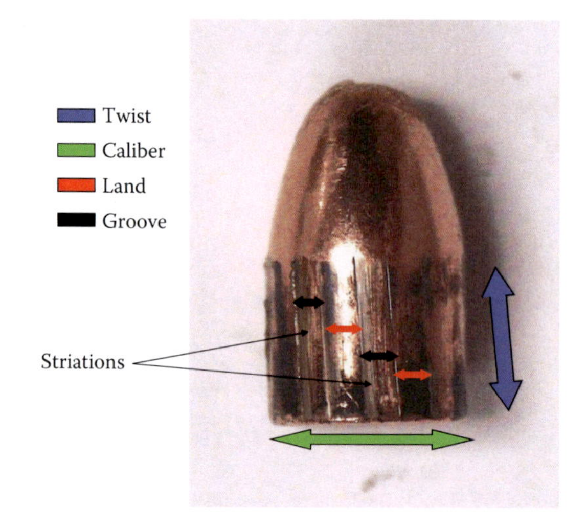

Figure 12.2
Class and individual characteristics of a fired bullet. (Courtesy of Chris Monturo, Precision Forensic Testing, Dayton, OH, http://www.precisionforensictesting.com.)

also the diameter of its base. When measuring the diameter in inches or millimeters, the following chart will help convert to the corresponding caliber.

Gun Caliber—Converting Measurements to Caliber

Gun Barrel Caliber	English System Measurement	Metric System Measurement
22 caliber	0.22 in.	5.56 mm
25 caliber	0.25 in.	6.35 mm
30 caliber	0.30 in.	7.62 mm
32 caliber	0.32 in.	7.65 mm
38 caliber	0.38 in.	9.00 mm
9 caliber	0.38 in.	9.00 mm
40 caliber	0.40 in.	10.0 mm
45 caliber	0.45 in.	11.25 mm

There are various types of bullets. There are many variations including hollow points, Teflon coated (armor piercing), and exploding bullets. See Figure 12.3 for typical bullet types, size, and weight.

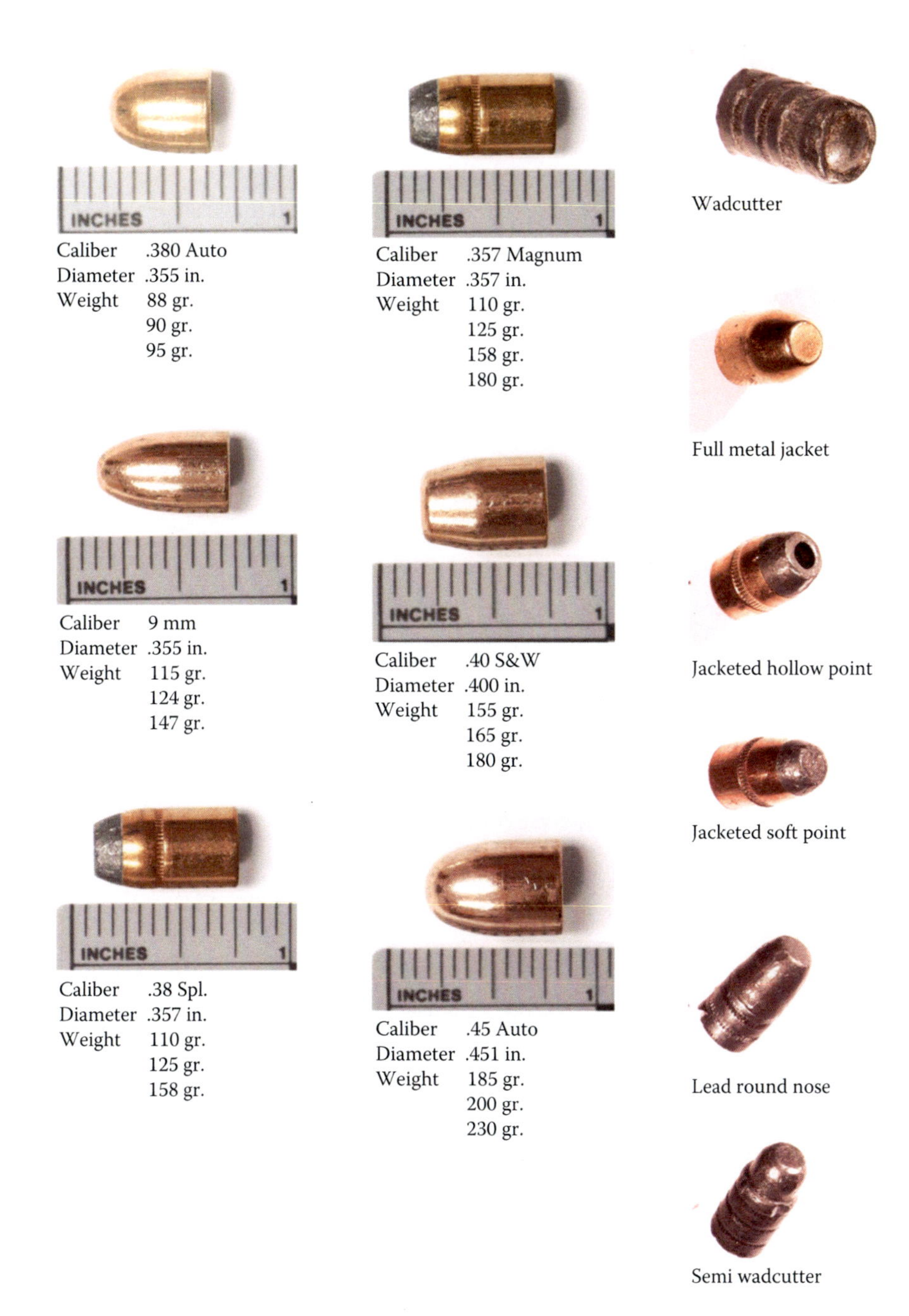

Figure 12.3
Bullet reference chart.

12.1.2 Cartridge Case Identification

Cartridges cases can yield similar information as bullets. There are a number of markings on cartridge cases that firearms examiners use for identification purposes. Marks that are important are firing pin impressions, extractor and ejector markings (except in revolvers), breech block markings, and sometimes, chamber markings.

When a bullet is fired, the cartridge case recoils back toward the shooter. A block of metal, called the *breech*, stops the cartridge case from hitting the shooter. This block contains markings that are transferred to the surface of the head of the casing. A comparison of breech block markings on cartridge casings is shown in Figure 12.4a.

After the bullet is fired from the cartridge, the casing must be removed, so that another round can be loaded into the gun. The metal *extractor* grabs the cartridge, so that it can be expelled from the chamber by the metal *ejector*. Extractor marks can be found on the lip of the cartridge case and extractor marks on the headstamp. Examples of both markings are shown in Figure 12.4b.

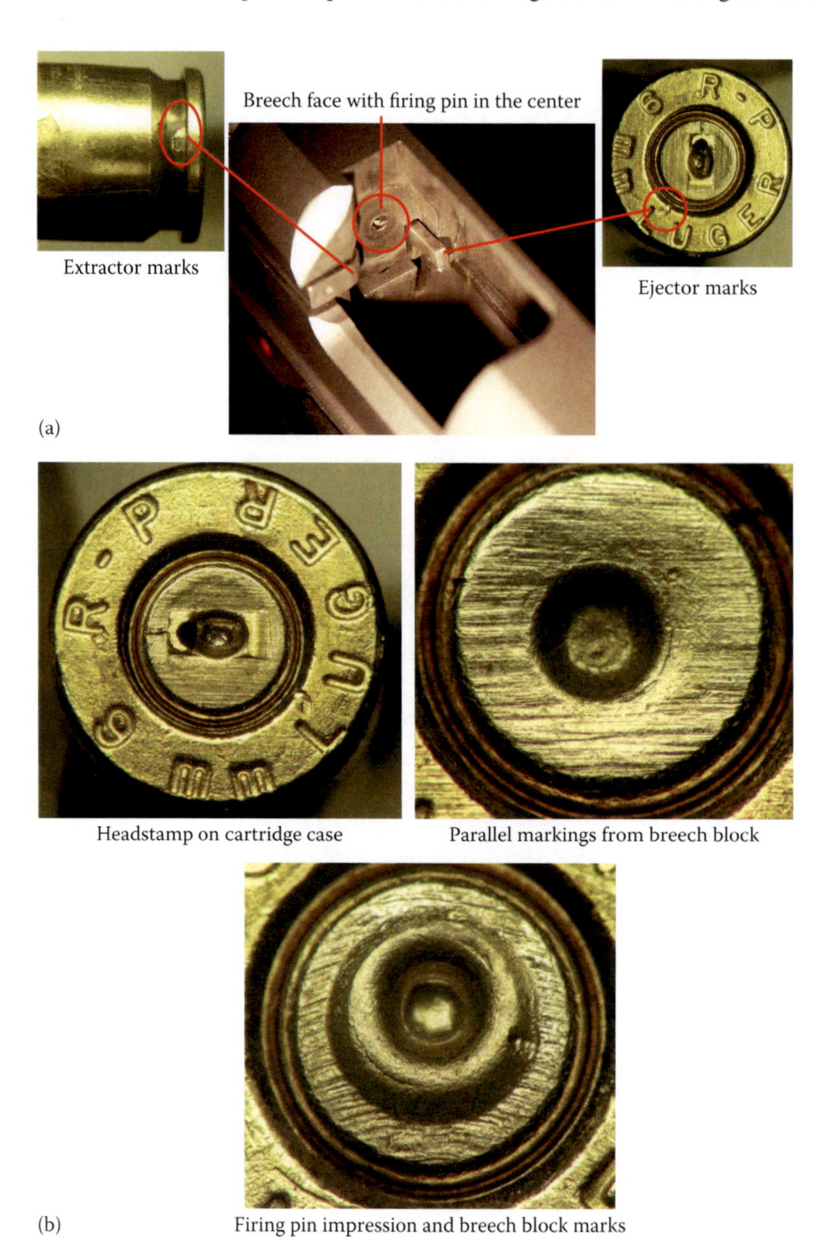

Figure 12.4
(a) Cartridge case markings after a gun is fired. (b) Chamber of a gun showing the breech block, firing pin, extractor, and ejector. (Courtesy of Chris Monturo, Precision Forensic Testing, Dayton, OH, http://www.precisionforensictesting.com.)

Striations on the sides of the cartridge case called *chamber marks* aid in identification. Chamber marks are produced by the expansion of gases and heating when the cartridge is fired. The casing is pressed tightly against the gun chamber and marks are impressed as it expands and moves. Figure 12.5 shows chamber marks on a cartridge case and the portion of the gun barrel that made the marks.

Cartridge cases are made of brass, nickel-plated brass, or aluminum. See Figure 12.6.

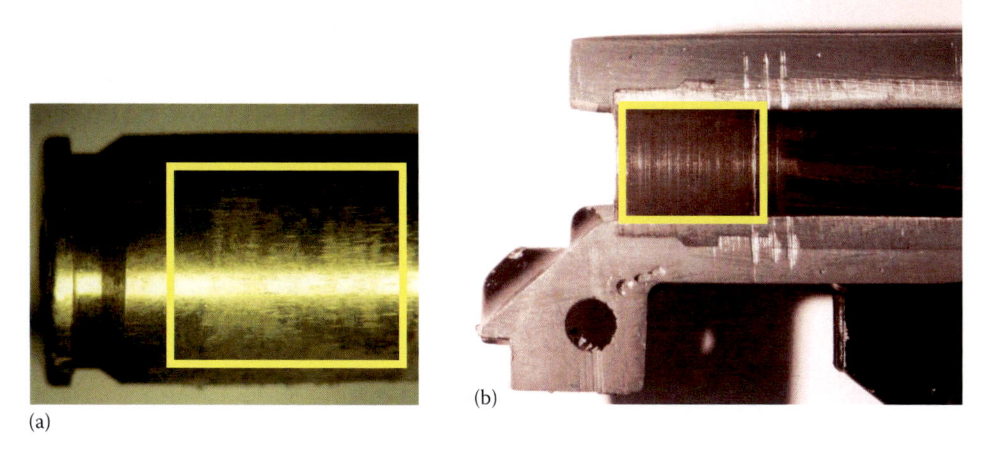

Figure 12.5
(a) Chamber marks on a fired cartridge case and (b) the gun barrel chamber. These fine striations can aid in identification. (Courtesy of Chris Monturo, Precision Forensic Testing, Dayton, OH, http://www.precisionforensictesting.com.)

Figure 12.6
Cartridge cases: (from left to right) aluminum, brass, and nickel-plated brass. (Courtesy of Chris Monturo, Precision Forensic Testing, Dayton, OH, http://www.precisionforensictesting.com.)

12.2 Pre-Laboratory Questions

1. What are the class characteristics that a firearms examiner uses in bullet identification?

2. Are there any individual characteristics that can match bullets? If so, what are they?

3. What characteristics are used in cartridge case identification?

4. Where can the examiner find individual striations that will match a crime scene cartridge case to a test fired case?

12.3 Scenario

There was a robbery at the local convenience store late at night by a lone gunman. The perpetrator fired three shots, one went into the wall behind the cash register and two were recovered from the store clerk who was injured in the hold up. All three bullets and their cartridge casings are in the lab for processing. A handgun was found two hours later by authorities hidden in the bushes behind a trash dumpster a block away. The handgun was test fired and those rounds with casings are in the lab for processing.

12.4 Materials

- Crime scene bullets and test fire bullets
- Crime scene cartridge casings and test fire casings
- Caliper or metric ruler
- Stereo microscope or hand lens
- Washable marker

12.5 Procedure

12.5.1 Bullet Identification

1. Measure the caliber of the bullet using the caliper or metric ruler. Caliber is determined by measuring the diameter of the base of the bullet (see Figure 12.2 for reference). Measure the caliber in millimeters. The metric measurement can be converted to inches by dividing the millimeter measurement by 25.4, since there are 25.4 mm in 1 in. Use the gun caliber reference chart to check the caliber. Place your results in Table 12.1.

2. Place the base of the bullet on the table with the tip upward. Look at the lands and grooves in the bullet and determine whether they twist to the right or left. Place your results in Table 12.1.
3. Place a starting mark on one of the grooves with the washable marker. Count the number of grooves and put the results in the table (the number of lands = the number of grooves).
4. Use the calipers or ruler to measure the width of a land and the width of a groove. Record the results.
5. Determine the type of bullet and record it in Table 12.1.

12.5.2 Cartridge Case Identification

1. Measure the caliber of the casing using the caliper or metric ruler. Caliber is determined by measuring the diameter of the base where the headstamp is located. Measure the caliber in millimeters. The metric measurement can be converted to inches by dividing the millimeter measurement by 25.4, since there are 25.4 mm in 1 in. Place your results in Table 12.2.
2. Record any headstamp markings in the table.
3. Look for lines on the area around the center of the head of the casing, where the firing pin hits. You may need to use the microscope or hand lens to see them. These are breech block lines. Are they straight parallel lines, are they arched parallel lines, circular lines or absent? Record these markings by sketching them in the table.
4. Sketch the shape and position of the firing pin impression on the casing.
5. Using the microscope or hand lens look for ejector marks on the headstamp area of the cartridge case. Note in the table if present and sketch the position.
6. Look for an extractor mark on the side of the casing lip. Note if present and sketch.

12.6 Follow-Up Questions

1. Which bullets, if any, have matching class characteristics? List the bullets and the matching traits.
2. Which cartridge casings, if any, have matching class characteristics? List the casings and the matching traits.
3. Soft point bullets and hollow point bullets have exposed lead which damages easily when it enters a target. How would this affect bullet identification?
4. Many criminals ream out a gun barrel to destroy the manufacturer's land and groove arrangement. What possible effect(s) could this have on the gun and on the bullet markings?

12.7 Basic Firearms Identification Worksheets

TABLE 12.1
Bullet Characteristics for Identification

Bullet Number	Caliber	Twist	Number Lands/ Grooves	Width of Land	Width of Groove	Type of Bullet
1						
2						
3						
4						
5						
6						
7						
8						

TABLE 12.2
Cartridge Case Characteristics for Identification

Cartridge Number	Caliber	Headstamp Marking (sketch)	Breech Block Lines (sketch)	Firing Pin Impression (sketch)	Ejector Mark (sketch)	Extractor Mark (sketch)
1						
2						
3						
4						
5						
6						
7						
8						

Lab 13

Basic Toolmark Identification

Screwdriver Comparison

13.1 Introduction

A toolmark is a scratch or other microscopic marking left by the action of a tool on an object. The important aspect of these marks is that no two toolmarks, even those left by the same type of tool, are identical. This implies that, in general, toolmarks can be individualized to a specific tool. The criterion of a match of known and unknown toolmarks is that there are a sufficient number of similarities and no unexplainable differences.

Class characteristics used for identification are the size of the working edge of the tool, the shape of the edge of the tool, the presence or absence of rib marks and the thickness of the tool edge (Figure 13.1).

Individual characteristics are the marks or imperfections in the tool caused by manufacture or in most cases due to use. Over time, a screwdriver tip gets uneven and scratched due to use. There may be pits or nicks on the tip or on the blade. See Figure 13.2 for examples of individual marks unique to a given screwdriver.

13.2 Pre-Laboratory Questions

1. What is the definition of a toolmark?

2. What characteristics make one screwdriver different from another?

3. A manufacturer makes 100 screwdrivers for distribution to the public. Are they all exactly alike? How are they the same? How are they different?

13.3 Scenario

The back door of a residence was pried open to gain entrance while the homeowners were on vacation. A small angular tool was used to pry open the door jam and the door knob plate, leaving an impression in the metal plate. The plate was removed and is in the lab for processing. Various tools were collected for comparison and are also in the lab.

Figure 13.1
Class characteristics of different screwdrivers: shape, size, and ribbing. (Courtesy of Chris Monturo, Precision Forensic Testing, Dayton, OH, http://www.precisionforensictesting.com.)

Figure 13.2
Individual characteristics of a screwdriver head. (Courtesy of Chris Monturo, Precision Forensic Testing, Dayton, OH, http://www.precisionforensictesting.com.)

13.4 Materials

- Six to eight different screwdrivers (suspect screwdrivers)
- Evidence from the break in—toolmark in metal plate
- Heavy-duty aluminum foil or can pieces (cut into 3″ squares)
- Casting material or modeling clay
- Hand lenses or stereo microscopes
- *Optional*: Digital camera or digital microscope

13.5 Procedure

1. Make a cast of the entire tip (top, bottom, and edge) of each screwdriver using the casting material.
2. If using Micro-Sil™ or ForenSil, follow the manufacturer's directions.
3. If using modeling clay, take a small piece of clay and smoothing it out into a 2″ square. Gently press one side of the tip of the screwdriver into the clay, leaving room for an impression of the other side of the tip.
4. Flip the tip over and make a second impression in the clay piece. Additionally, make an impression of the end of the screwdriver tip. Your clay piece should look *something* like the diagram below.

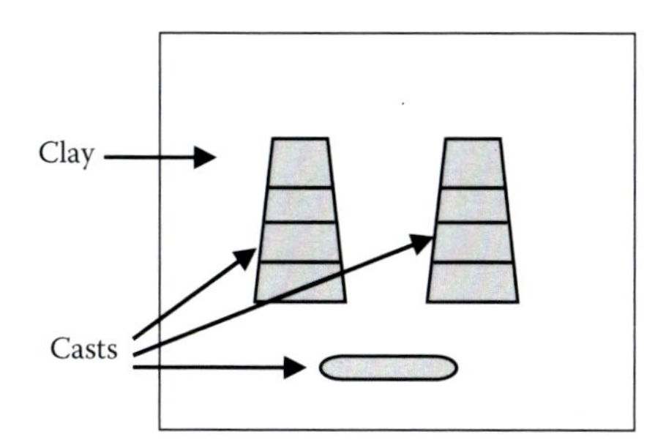

5. Label your cast. Repeat for all suspect screwdrivers.
6. Using your suspect casts, determine which matches the cast of the evidence. Be prepared to support your claim with details about the match.
7. Using the same screwdrivers, gently scrape them across the aluminum foil sheet, using both sides of the tip. Label your suspect scrapes.
8. Compare your test scrapes to the evidence. Determine which matches the cast of the evidence. Be prepared to support your claim with details about the match.
9. If available, take a digital photograph of the evidence and suspect casts and scrapes. Use the photographs for a comparison match of the evidence (see Figure 13.3).

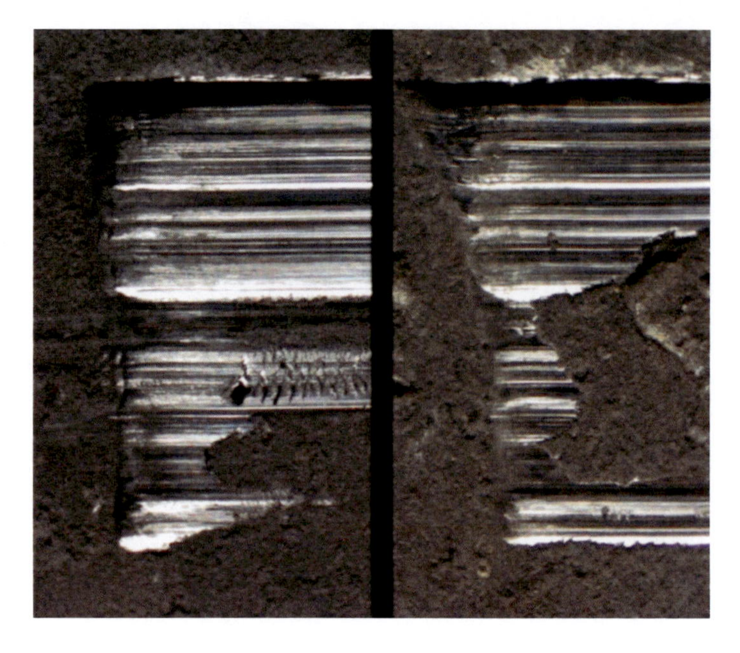

Figure 13.3
Comparison of scrape marks. (Courtesy of Chris Monturo, Precision Forensic Testing, Dayton, OH, http://www.precisionforensictesting.com.)

13.6 Post-Laboratory Questions

1. Based on the lab tests, which screwdriver was used in the home invasion?

2. What characteristics helped in making the identification of the tool used in the crime?

3. How are individual characteristics different from class characteristics when classifying toolmarks?

Lab 14

Shoeprints

14.1 Introduction

There are many different types of impressions that may be found at a crime scene. Evidence can be identified by grouping items according to class characteristics. Class characteristics are general distinctions that can be made about a specific item. Individual characteristics are needed to uniquely individualize evidence. These types of characteristics are unique to the shoe, meaning that no other shoe has these specific characteristics. This puts the shoe into a group all alone. These types of characteristics may be a product of wear, use, environment, etc.

Footwear may be found near entrances and exits of the space where the crime was committed. This type of evidence can be used to identify or exclude possible suspects. Unique characteristics in the impression are used to identify the shoe of the wearer.

By determining the type of impression or print found at the crime scene, distinguishing between a positive and negative impression may also be useful. Positive impressions or prints are when a clean surface has an addition of an impression, such as a muddy footprint left on a clean floor. A negative impression or print is when a dirty surface has a removal of an impression, such as a footprint impression made on a dirty surface. This can help to determine the order of events in a location, which can be helpful to the investigator.

14.2 Pre-Laboratory Questions

1. How do individual characteristics originate?

2. What are positive and negative impressions?

14.3 Scenario

Zachary Stevens' house was burglarized early Friday morning. The police officers arrived on the scene and examined the points of entry into the house. Mr. Stevens claimed that there were several items of great value in the house that had been stolen, with muddy shoeprints tracked throughout his house. A muddy shoeprint impression was also found outside in the mud. A cast of the footprints was done and was need to be compared to the suspects' shoes.

14.4 Materials

- Dry dirt in a cardboard box
- Dental stone mix and water
- Shoes
- Aluminum foil
- Large ink pad
- White cardstock paper

14.5 Procedure

1. Take a 3D footwear impression
   a. Take the box filled with dirt and step into it to create a shoe impression.
   b. Spray the impression with hairspray about 8 in. away from soil in sweeping motion.
   c. Mix plaster of Paris in the ziplock bag with enough water to produce a pancake batter-like consistency.
   d. Use aluminum strips to make a border around the print.
   e. Pour evenly over the footprint. Allow time to harden (may take 1-2 hours). Label the plaster with date and name before it dries.
   f. Remove any dirt from the print once it dries and examine the print for individual characteristics.
2. Take a 2D impression of the footwear
   a. Ink the bottom of the same shoe with the large ink pad.
   b. Press inked impression of the shoe onto the white cardstock.
   c. Make sure to evenly apply pressure from heel to toe.
3. Compare the impression
   a. Find similar features within the 2D impression and the 3D impressions.

14.6 Follow-Up Questions

1. You can have two separate items with the same class characteristics while having different individual characteristics. That being said, can you have two separate items with different class characteristics while having the same individual characteristics? Why/how?

2. Was the footwear cast that you created a positive or negative impression?

3. What were the similarities and differences between your 2D shoe print and 3D shoe impression?

14.7 Shoeprints Worksheet

2D Footwear Impression
Describe what class characteristics observed on the impression:
Describe what individual characteristics observed on the impression:
3D Footwear Impression
Describe what class characteristics observed on the impression:
Describe what individual characteristics observed on the impression:
Final Analysis
Could the 2D and 3D impressions have come from the same shoe? How do you know?

Unit 3

Forensic Biology

Lab 15

Serology

15.1 Introduction

The science that deals with the properties and reactions of different bodily serums is known as *serology*. It is used in forensic science to investigate different bodily fluids, which may be found at a crime scene. Serology provides a quick, efficient, and an inexpensive way to determine if a stain had originated from an organism. This type of evidence can also provide context to the crime scene by identifying which bodily fluid is present. Identification of the type of biological serum present in a stain can offer insight as to the type of crime, which may have been committed. These bodily fluids can include blood, urine, semen, and saliva.

When examining bodily fluids, a presumptive and a confirmatory test can be used to test the sample that is in question. A presumptive test provides a quick and easy examination, either at the scene or in the laboratory, to show the possibility of a particular substance. A confirmatory test provides a definitive result for the substance that is in question.

The presumptive tests used in this laboratory include the Kastle-Meyer or phenolphthalein test for blood, the starch-iodine test for saliva, and ultraviolet light to detect urine. The Kastle-Meyer test produces an immediate pink color when the solution reacts to the blood due to its pH level. The starch-iodine test detects the presence of human saliva through the breakdown of starch into simple sugars if saliva is present. For urine, the presence of a specific compound, urobilin, can be seen by resulting in a bright yellow fluorescence under ultraviolet light.

15.2 Pre-Laboratory Questions

1. What does serology study?

2. What is a presumptive test for blood?

3. True or false: A confirmatory test is specific for a particular substance.

15.3 Scenario

On November 23, at 3:27 p.m., the police officers arrived on the scene of a reported violent attack. The victim, Kristina, was found lying on the living room floor of her house tangled up in a bed sheet. She was severely bruised and had bite marks on her neck. There were wet spots on the sheet that was around her. The officers noticed red stains on the floor coming from the bedroom. At the hospital, the doctor swabbed the bite marks on Kristina's neck to obtain possible saliva from the perpetrator. The forensic serologists also examined the sheet and collected the red stains.

15.4 Materials

- Cotton swabs
- Plastic tubes (2ml)
- Glass microscope slides
- Testing solutions
- Positive and Negative Samples

15.5 Laboratory Procedures

15.5.1 Basic Laboratory Procedure

1. Presumptive tests
   a. Saliva: starch-iodine test
      i. Place a cotton swab in your mouth and swab the inside of your mouth.
      ii. Drop the cotton end of the swab in a microcentrifuge tube and break off the stem of the stick.
      iii. Add three drops of the starch solution.
      iv. Do the same with a negative control—a dry cotton swab.
      v. Incubate for 1 h at 37°C.
      vi. Add two drops of the iodine solution and note immediate color change. (*Note:* no color change is a positive result—because amylases break down starch.)
   b. Blood: Kastle-Meyer test
      i. Take two cotton swabs to use as your controls. For the positive control, add a drop of the known blood to one of the swabs. For the negative control, add a drop of distilled water to the other swab.
      ii. Apply two to three drops of 95% ethanol, followed by two to three drops of phenolphthalein. Wait approximately for 10 seconds, and then add two to three drops of 3% hydrogen peroxide to each cotton swab. (An immediate pink color change with the addition of hydrogen peroxide is a positive result. A false positive may result if no time is given between drops of phenolphthalein and hydrogen peroxide.)
      iii. The positive control should turn pink and the negative control should not change color at all. A positive presumptive test for blood will change color to pink.
      iv. Obtain three evidence stains. Using three different swabs, moistened with distilled water and roll one over each stain.
      v. Follow step ii above on each swab to determine if the stain presumptively tests positive for blood.
      vi. Record your results of the test in the chart.
   c. Urine: ultraviolet lamp test
      i. Obtain a cotton swab and remove it from the packaging.
      ii. Locate your nearest bathroom and swab a toilet seat, toilet lid, or a toilet lip for urine.

iii. After returning to the classroom, take an additional cotton swab and wet with water. This will act as a negative control for your test.

iv. Take both cotton swabs to the ultraviolet light source and display them underneath the light, a yellow fluorescence under the ultraviolet light indicates a positive presumptive presence of urine.

v. Record your results.

15.5.2 Advanced Laboratory Procedure

1. Sperm confirmatory test
   a. Christmas Tree stain
      i. Obtain a prepared dried semen microscope slide.
      ii. Place three drops of Christmas Tree stain A on the dried semen stain and allow it to sit for 15 minutes.
      iii. Gently rinse the slide with distilled water to remove stain A.
      iv. Place three drops of Christmas Tree stain B on the dried semen stain and allow it to sit for 10 seconds.
      v. Gently rinse the slide with 95% ethanol to remove stain B.
      vi. Allow slide to air dry.
      vii. Apply one or two drops of immersion oil to the sample slide.
      viii. Examine the slide using a compound microscope with the 100× objective. Identify the sperm. See Figure 15.1 for examples of Christmas Tree stain of sperm cells.

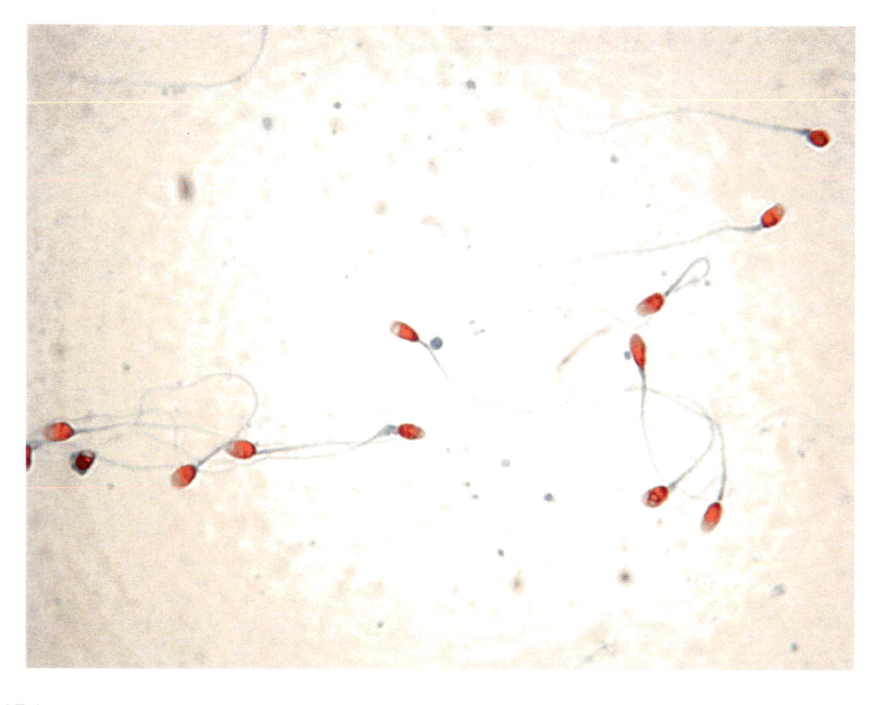

Figure 15.1
Christmas Tree stain of sperm cells.

15.6 Follow-Up Questions

15.6.1 Basic Follow-Up Questions

1. Think back when you were conducting your preliminary test for blood. If the known blood sample did not turn your swab pink, would that be considered a false positive or false negative? Why?

2. Using what you know about how the Kastle-Meyer test works to detect the presence of blood, how could a substance that is not blood create a false positive for the test?

3. If the swab you tested for urine responded the same way the positive control did, what could you conclude?

15.6.2 Advanced Follow-Up Questions

1. After conducting the confirmatory test for sperm, what colors did the sperm turn?

2. An examiner performs the Christmas Tree stain on a stain, which is suspected to be semen. Upon observation under the microscope, no colored spots are observed and the examiner concludes that the stain was not semen. Is this the correct conclusion? Why?

15.7 Serology Worksheet

Kastle-Meyer Test
Positive control results:
Negative control results:
Stain A results:
Stain B results:
Stain C results:
Saliva Test
Positive control results:

(*Continued*)

Negative control results:
Urine Ultraviolet Lamp Test
Positive control results:
Negative control results:
Sperm Confirmatory Test

Color of head:	Draw what you see:
Color of tail:	
Other observations:	

Lab 16

Pathology

The Autopsy

16.1 Introduction

One of the most critical parts of crime scene investigation involving death is the autopsy. When most people hear the word *autopsy*, they think dissection and looking inside the human body. However, it is just as important to carefully examine the exterior of the body.

The term *autopsy* means to *see with one's own eyes*. A pathologist is the medical professional who carries out an autopsy with assistance from other personnel. A forensic pathologist specializes in autopsies in which the cause and manner of death is questionable. The pathologist carries out two through examinations: examining the outside of the body or the *external exam* and examining the inside of the body or the *internal exam*. Figure 16.1 shows the autopsy of a pickle.

The external examination of the body is just as crucial as the internal exam. It can give clues about the cause and manner of death, locate identifying markings such as tattoos or unusual moles or body markings, and can yield trace evidence (hairs, fibers, etc.) on the body that can help associate the deceased with possible suspects. The body is extensively photographed. Bruising, wounds, and trauma to the body are noted as well as entry and exit gunshot wounds or cuts.

The internal examination is started after the external clues are carefully documented. A standard Y-shaped incision is made on the ventral side (chest area) of the body. The top edges of the Y represent the cuts at the two shoulders, the Y comes together at the mid-chest, and the single tail of the Y carries down to and around the naval into the groin area. The skin is then reflected back and the internal cavities examined.

Body fluid samples including blood, urine, and other fluids are usually removed and sent to the toxicology lab to determine if there are drugs or poisons in the body that could have caused or contributed to death. Major organs are removed, weighed, and measured. Body organs are also examined for disease or damage. Wounds or injuries that appeared in the external exam are traced. Bullets or other foreign material are located, photographed, and removed. These materials would be tagged as evidence and sent to the crime lab for analysis. The body may be X-rayed for comparison to any X-rays taken of the victim prior to death. This process helps to identify or to verify the identity of the deceased.

Once all the examination data is taken and lab results analyzed, the pathologist makes a determination of the cause of death. Post autopsy, the Y incision is sutured and the body released for burial or cremation. Figure 16.1 shows the autopsy of a pickle.

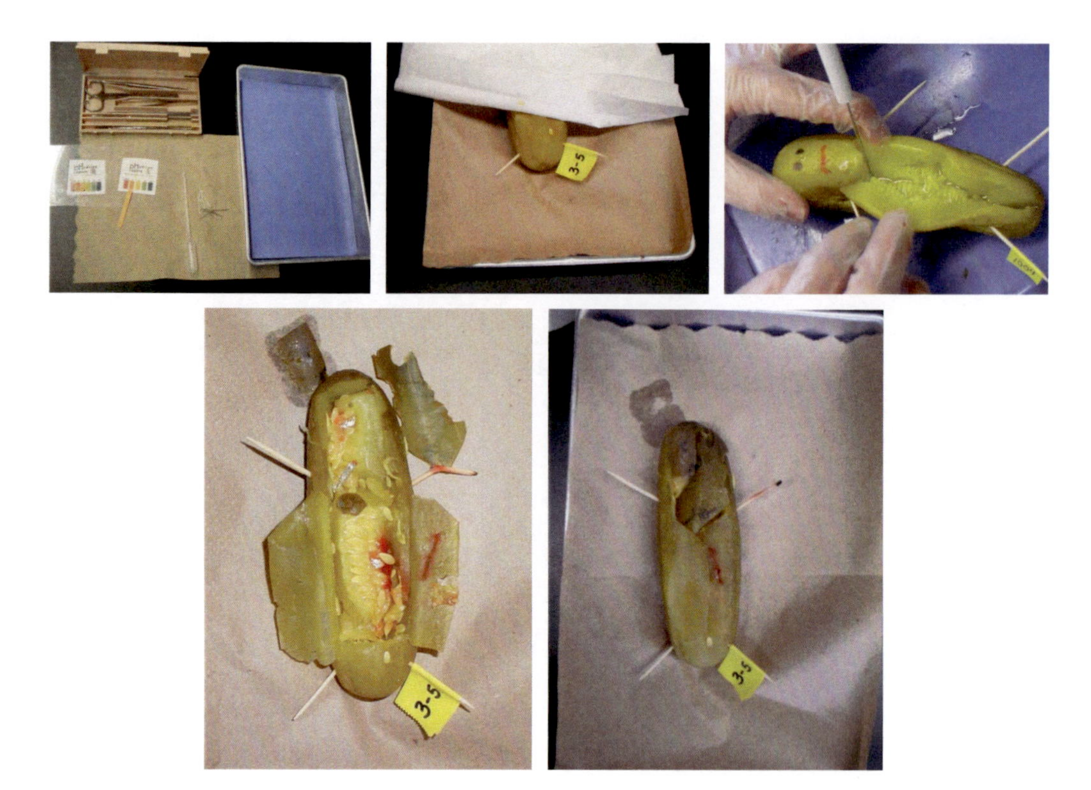

Figure 16.1
Pickle autopsy. (Courtesy of Kathy Mirakovits, Forensic Science Education Consulting, LLC.)

16.2 Pre-Laboratory Questions

1. How is a forensic pathologist different from a pathologist? How are they alike?

2. In the external examination, what types of evidence should be noted?

3. What types of evidence can be noted during the internal examination?

16.3 Materials

- Dissecting pan
- Dissecting instruments
- Victim

- Evidence collection materials
- Camera (if autopsy is to be photographed)
- Pipette for fluid collection
- pH paper
- Gloves
- Goggles

16.4 Procedure

Note: Use the autopsy report sheet to make all notations and sketches.

1. Do a thorough external examination of the body of the deceased, noting any markings or trauma to the body.
2. Using the scalpel, make the Y incision, starting at the top of the chest area between the shoulder and neck on each side of the body. The two cuts should meet at the sternum in the middle of the chest. Continue the incision down the middle of the body and around the navel ending in the groin area.
3. Reflect back the skin to expose the organs.
4. Remove or cut open the rib cage to expose the thoracic cavity, which includes the heart and lungs.
5. Complete the internal examination, noting any trauma to the major organs of the body.
6. After the examination is complete return any organs that were removed for examination back to the body. Suture the Y incision if requested by your instructor.
7. Complete the autopsy report by making a determination of the cause of death.

16.5 Follow-Up Questions

1. Why is it important to do a thorough external examination?
2. How do X-rays assist in the autopsy?
3. The deceased was found in the water. Would this have an effect on the autopsy?
4. What role(s) does the autopsy play in a crime investigation?
5. In the United States, the person authorized to perform the forensic autopsy can be either a coroner or a medical examiner. Research the difference. Is there any controversy involved? Comment on who performs autopsies in your location.

16.6 Pathology Worksheet

Lab 16—Autopsy Report Form

Pathologist ______________________ Assisted by ______________________ Date___________
Deceased Name ____________________ Tag Number ________________ Case # _________

EXTERNAL EXAMINATION

Examine the ventral (front) and dorsal (back) sides of the decedent. Note any marks on the diagrams below and label.

VENTRAL VIEW

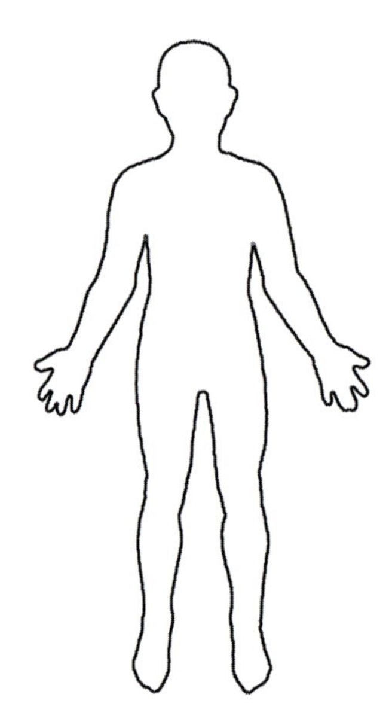

DORSAL VIEW

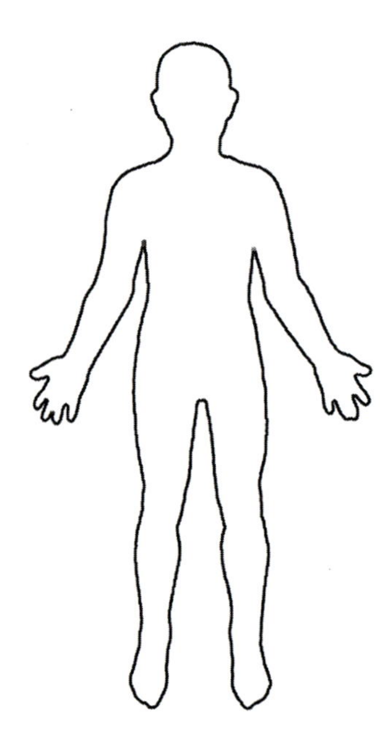

INTERNAL EXAMINATION

- Make the Y incision and cut away the skin from the chest cavity.
- Note any trauma or would tracks.
- Cut away the rib cage either by cutting through the sternum and spreading the rib cage or by making two vertical cuts through the right and left ribs and removing the rib cage.
- Examine the thoracic (chest) organs.
- Examine the abdominal organs.
- Collect any foreign matter and tag for the crime lab.
- Collect body fluids and tag for the toxicology lab.
- Sketch the chest cavity and abdominal cavities, noting any findings.

THORACIC VIEW

ABDOMINAL VIEW

OPTIONAL: BRAIN EXAMINATION

- Remove the top of the skull carefully (brain is very soft and gets damaged easily).
- Run the probe around the skull cavity to loosen the brain.
- Gently remove the entire brain and examine for hemorrhaging and discoloration.
- *Notes:*

OPTIONAL: TOXICOLOGY TEST

- Collect a sample of body fluid using the pipette. Test the pH of body fluid using pH test paper.
- Resulting pH = _________. Is the body fluid acidic, basic, or neutral?
- Normal pH of this body is _________.

CONCLUSION

What is your finding about cause of death of this patient? Support your opinion with specific details from the autopsy.

Signature ____________________________________ Date ____________________

Lab

DNA Analysis

17.1 Introduction

Deoxyribonucleic acid (DNA) is found in all cells with the exception of red blood cells. The structure of the DNA consists of a backbone composed of alternating sugar molecules and phosphates with one of four nucleotides attached to each sugar molecule: adenine (A), guanine (G), cytosine (C), and thymine (T). The characteristic double helical structure of the DNA results from hydrogen bonding of the bases of the nucleic acids of the DNA. The bonding in DNA is highly specific, in that T only bonds with A and G only bonds with C. Repeating units of these base pair sequences compose the sequence of DNA. The order of the sequencing of base pairs in DNA is determined by genetics and serves as the genetic code, which determines many characteristics of a person.

Prior to DNA analysis, biological material was characterized into a specific group to narrow the possible origin. ABO blood typing was used to classify blood based on the antigens present on the red blood cells. A person's blood is classified as A, B, AB, or O corresponding to blood cells with antigen A, B, A and B, or none, respectively. Each blood type has an antibody that opposes the antigen not present in the red blood cells for the specific blood group. For example, type A blood possesses the A antigen and the anti-B antibody. The blood type of a specific blood sample can be determined by mixing a blood antibody and antigen and when the same type of antibody and antigen are combined, the blood will clump together. Therefore, when type A blood is exposed to anti-A antibody, the blood will clump. This explains why people with type A blood cannot receive transfusions of type B blood and vice versa. The table below summarizes the blood groups in the ABO system and their corresponding antigens and antibodies.

Blood Group	Antigen	Antibody
A	A	anti-B
B	B	anti-A
AB	A and B	none
O	none	anti-A and anti-B

Overall, this makes persons with type O blood universal donors, meaning that this blood can be accepted by any other blood type due to the absence of ABO antigens. Similarly, individuals with type AB blood are universal recipients due to the absence of ABO antibodies and can therefore accept any other blood type in a transfusion.

17.2 Pre-Laboratory Questions

1. Name the four base pairs found in DNA. Identify which other base pair(s) bond with each one.

2. What are some common sources in which DNA can be found?

3. What type of blood type must you have to be considered a universal receiver? Why?

17.3 Scenario

Over the past couple of weeks, several bodies have been discovered and determined by the medical examiner to have been killed in the same manner of death. It is in question if the same person has committed the murders. However, police officers noticed each of the victims strangely had strawberries present at the crime scene. Special investigators suspect this strange piece of evidence to possibly be part of the killer's signature, and have dubbed the unknown suspect as the *Strawberry Killer.* After the strawberries were collected and preserved, investigators are requesting DNA analysis to be conducted to determine if the strawberries from the different crime scenes could have come from the same strawberry patch. It is your job as a DNA analyst to extract the strawberry DNA for future analysis.

17.4 Materials

- Ziplock bags
- Paper clips
- Strawberries
- Aluminum foil
- Plastic tubes with closure (10ml)
- Ice
- Extraction solution

17.5 Laboratory Procedures

17.5.1 Basic Laboratory Procedure

1. Strawberry DNA
   a. Place the 95% ethanol on ice.
   b. Place one or two strawberries in the ziplock bag.
   c. Mash up the strawberries for two minutes.
   d. Measure out 10 ml of the DNA extraction buffer and add to the bag to mash for two more minutes. Try not to create a lot of soap bubbles. Let it sit for two more minutes while performing the next step.

e. Place a funnel in a 50 ml conical vial and place it in an ice bucket. Place paper towel over the funnel. Pour the contents of the bag into the funnel to separate the strawberry mush from the juice. Let the juice drip into the 50 ml vial for approximately five minutes.

f. Gently squeeze the remaining liquid from the paper towel into the vial. You should have 8–10 ml of a pink liquid. Record the amount of liquid you have.

g. Carefully pipette twice the volume of your pink liquid (if you have 8 ml of pink liquid then add 16 ml 95% ethanol) of ice-cold 95% ethanol down the side of the 50 ml vial. Mix the vial very gently and this will still have a layer over the extract.

h. Keep on ice for two minutes. Then look for a fluffy white cloud of precipitate where the layers came together.

i. Use the paper clip hook to pull out the clear, glassy DNA.

j. Place the DNA on a piece of parafilm and examine it. Record what you observe.

17.5.2 Advanced Laboratory Procedure

1. ABO blood test

   a. Choose a finger to prick to draw blood from and wipe it with an alcohol wipe provided.

   b. Shake the hand of the finger chosen to prick around for about 30 seconds to increase blood circulation to the fingertips.

   c. Using the lancet provided, prick the chosen finger and gently squeeze the fingerpad to draw blood from the pricked area.

   d. Place a drop of blood on each side of the ABO slide guide card.

   e. On the left side labeled *anti-A serum* place one drop of *A blood grouping serum* (blue solution).

   f. On the right side labeled *anti-B serum* place one drop of *B blood grouping serum* (yellow solution).

   g. Using a toothpick, mix the drop of blood with the appropriate serum separately by combining the two drops using a circular motion to bring the two drops together. Use a different toothpick to combine the different blood/serum samples.

   h. Record any clumping between the serum and blood mixture and determine your blood type from the key to observations provided on the ABO slide guide.

17.6 Follow-Up Questions

17.6.1 Basic Follow-Up Questions

1. What did you observe when you added the ethanol to the strawberry liquid solution?

2. You received a small sample of blood, too small for proper DNA analysis. You used short tandem repeats (STR) amplification in order to obtain enough DNA for analysis. Later on, you found out your sample was collected out of a public bathroom, which might have contaminated your already small sample with DNA from other people. What is a risk associated with DNA analysis that could now affect your results.

3. A student conducts this lab but forgets to add the DNA extraction buffer before straining the mashed strawberries and adding the ethanol to extract the DNA. What would be the expected result upon the addition of the ethanol to the strawberry liquid?

4. When isolating the strawberry DNA, what was the purpose of the detergent and the ethanol?

17.6.2 Advanced Follow-Up Questions

1. What was your blood type? What was your partner's blood type? What types of blood can both you and your partner receive?

17.7 DNA Analysis Worksheet

Describe your strawberry DNA.

Describe what you see when performing your ABO blood test:

Anti-A | Blood | Anti-B

(Clumping/No Clumping?)	(Clumping/No Clumping)

	Anti-A	Anti-B	Blood Type
Observable Result Options	Clumping	No Clumping	A
	No Clumping	Clumping	B
	No Clumping	No Clumping	O
	Clumping	Clumping	AB

What is your blood type?	
What is your partner's blood type?	

Lab 18

Hairs

18.1 Introduction

Hair analysis is circumstantial; however, hair is a common piece of evidence retrieved from crime scenes. Hair analysis is conducted by two different methods: by observation of the macroscopic and microscopic characteristics of the hair. Macroscopic hair analysis inspects the overall appearance of the hair. Macroscopically, a hair strand contains three different sections: a root, a shaft, and a tip. Each of these sections reveal different context concerning the hair's origin. The root can be examined to assess the growth stage of a hair. Hair is considered to be in one of three growth stages, the *anagen*, the *catagen*, or the *telogen* stage (Figure 18.1). The anagen stage is the phase in which the hair is actively growing. The catagen phase is considered the transitional phase since cell production shuts down. In the telogen phase, the hair rests in the follicle and is held in place by mechanical connection only through the root bulb sitting in the follicle. The root can also be examined for the presence of sheath material. This can offer context to determine whether a hair was forcibly removed and if DNA analysis can be conducted on the hair, as DNA cannot be extracted from the hair directly.

The shaft can be assessed to aid in body area determination of a hair. Head, facial, chest, and axillary hair (armpit) often have an even diameter throughout the length of the hair shaft; limb and eyebrow/eyelash hairs differ in diameter by expressing tapering of diameter, being thicker toward the root and thinner toward the tip. Overall, the macroscopic analysis of the hair is significant in assessing the context of the hair discovered as evidence, such as body origin or if it was naturally lost or forcibly removed, to provide context to the crime being investigated from which the hairs were involved.

The microscopic examination of a hair strand includes details that are not seen unless closely examined by a microscope; such as the anatomy of a hair. Hairs typically have three microscopic layers called the *cuticle*, *cortex*, and the *medulla*. An example of this is shown in Figure 18.2.

The cuticle is the outer protective coating of a hair, which grows in scale-like patterns. The cortex is the next layer underneath the cuticle, which contains the pigment of the hair. The center of a hair can contain a layer called the *medulla*. The medulla can be continuous, absent, interrupted, or even fragmented.

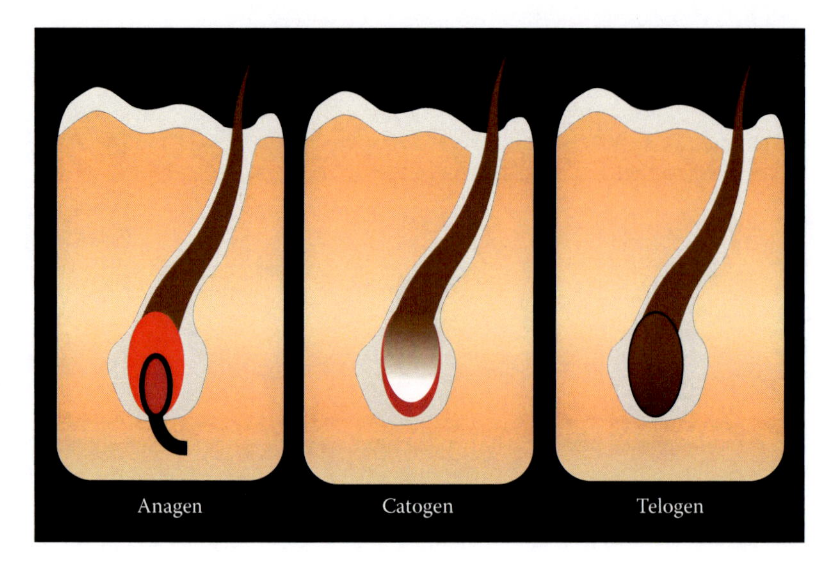

Figure 18.1
Growth stages of a hair.

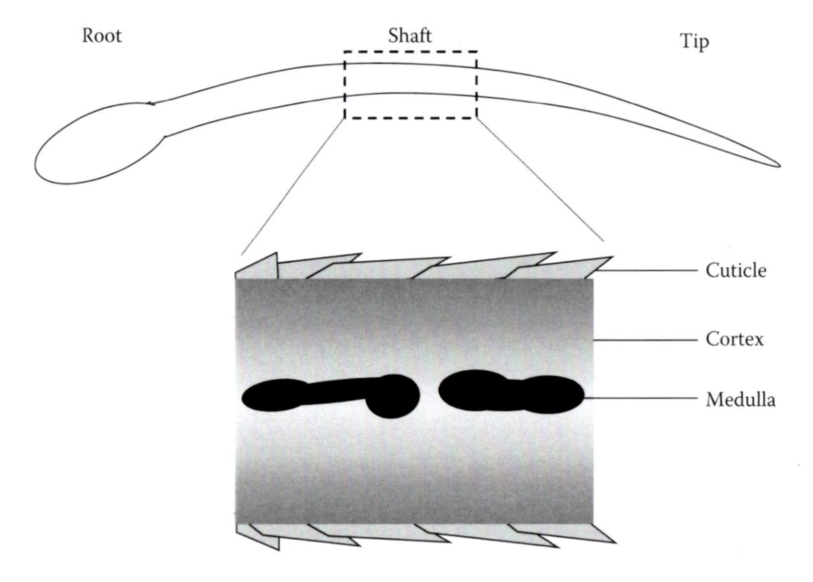

Figure 18.2
Macroscopic and microscopic analyses of the hair.

18.2 Pre-Laboratory Questions

1. What are the three macroscopic parts of the hair?
2. What are the three microscopic parts of the hair?
3. What part of the hair is needed in order to use it to test for DNA?

18.3 Scenario

A car was found out on a country road last night. After looking up the license plate number, it was found that the car had previously been reported as being stolen. It was examined and several types of hairs were found in the car. The hairs were collected and sent to the crime lab to determine if the hairs were of humans to help in identifying the possible car thief. The stage of growth that the hairs were in was also determined.

18.4 Materials

- Glass microscope slides
- Glycerin
- Glass cover slips
- Hair (human and animal) samples
- Clear nail polish

18.5 Procedure

1. Macroscopic feature examination using stereomicroscope
   a. Identify the root end and the tip end of the hair. Draw what the hair looks like.
   b. Identify what stage of growth the hair is in and the shape of the root.
   c. Identify characteristics such as color and shaft form and presence of banding.
   d. Measure the diameter and length of hair and record in centimeters.
   e. Repeat for each type of hair.

2. Microscopic examination using compound microscope
   a. Place the hair on a slide and apply a drop of glycerin to the hair. Place a coverslip over the hair and oil.
   b. Identify and record the type of cuticle, cortex and medulla width, medulla form, root characteristics, and pigmentation orientation.
   c. Examine the cross section of the three anthropological groups and note their differences.
   d. Repeat for each type of hair.
3. Scale casting
   a. Apply a thin coat of nail polish to a microscope slide.
   b. Position the human hair in the nail polish while it is still wet and let it dry completely.
   c. Gently remove the hair from the cast.
   d. Observe the scale pattern using the compound microscope. Draw what you see.
   e. Repeat steps a through d on an animal hair.

18.6 Follow-Up Questions

1. What were the three stages of hair growth? In which stage were your hairs?

2. What part of the hair were you looking at using the scale cast?

3. What characteristics could you use to distinguish a limb hair and a head hair?

18.7 Hairs Worksheet

SAMPLE		
Color		
Length		
Shaft form		
Diameter		
Cuticle		
Cortex		
Medulla		
Root shape		
Pigmentation		
Medullary index		

Lab 19

Forensic Anthropology

Determination of Stature

19.1 Introduction

A forensic anthropologist is the professional who is given the task of determining the age, gender, race, and stature of skeletal human remains, with the ultimate goal of identifying the individual. If enough bones and certain key bones are available (skull, pelvis, and long bones), the task can be uncomplicated. When only a few bones are available, the identification becomes more difficult and only a few traits of the decedent can be determined.

This activity concentrates on how measuring the length of the long bones of the body is used to estimate stature. The bones used by anthropologists to determine the approximate height are the bones of the arms and legs: the radius, ulna, humerus, tibia, fibula, and femur.

19.2 Pre-Laboratory Questions

1. What is the primary job of the forensic anthropologist?
2. What class characteristics (traits) would the forensic anthropologist be able to determine from a set of skeletal remains?
3. Brainstorm what other types of information could be gleaned from studying the skeletal remains of a body discovered at a potential crime scene.

19.3 Scenario

A deer hunter was walking toward his hunting blind deep into the forest off State Road 14, when he came upon some bones that looked human. He called the authorities and the bones were documented and collected. The bones collected were a partial skull with dentition, a left and right pelvis, a femur, and a humerus. The skull and pelvis were analyzed first and it was determined that the bones are of a female, possibly of African American descent. Age of the female is approximately 20–25 years. The height of the woman needs to be determined.

19.4 Materials

- Metric tape measure
- Calculator

19.5 Procedure

1. Two students must work together in this activity.
2. Measure the approximate length (in centimeters) of your partner's humerus, ulna, femur, and tibia.
3. Record your length values in Worksheet 19.7.
4. Using the regression formulas in Table 19.1, and the correct race and gender category, calculate an approximate height. A step-by-step example is shown below.

Example Problem

a. Caucasian male humerus = 36 cm
b. Insert length measurement into formula from Table 19.1: 2.89 (36 cm) + 78.10
c. Calculated height in centimeters = 182.14 cm.
d. Standard deviation (range of possible values) = +/−4.57.
e. Add 4.57 to calculated height of 182.14 cm = 186.71 cm, the maximum height.
f. Subtract 4.57 from calculated height of 182.14 cm = 177.57 cm, the minimum height.
g. Divide the range values (max and min) by 2.54, since there are 2.54 cm in an inch, and round to the nearest inch.
    i. Maximum height = 74 in
    ii. Minimum height = 70 in
h. Convert to feet and inches: Range is from 6′2″ to 5′10″
    i. Average height = 6′

5. Use the tape measure to determine the actual height of your partner and record it in Worksheet 19.7.
6. Calculate the percent difference between the long bone measurements and the actual height.

$$\text{Percent difference} = \frac{\text{Accepted} - \text{Experimental}}{\text{Accepted}} \times 100\%$$

TABLE 19.1
Stature for Males and Females, Various Ethnic Groups

Race/Sex	Formula (cm)	Standard Deviation
Caucasian male	2.89 * humerus + 78.10	±4.57
	3.79 * radius + 79.42	±4.66
	3.76 * ulna + 75.55	±4.72
	2.32 * femur + 65.53	±3.94
	2.42 * tibia + 81.93	±4.00
	2.60 * fibula + 3.86	±3.86
Caucasian female	3.36 * humerus + 57.97	±4.45
	4.74 * radius + 54.93	±4.24
	4.27 * ulna + 57.76	±4.30
	2.47 * femur + 54.10	±3.72
	2.90 * tibia + 61.53	±3.66
	2.93 * fibula + 59.61	±3.57
African male	2.88 * humerus + 75.48	±4.23
	3.32 * radius + 85.43	±4.57
	3.20 * ulna + 80.77	±4.74
	2.10 * femur + 72.22	±3.91
	2.19 * tibia + 85.36	±3.96
	2.34 * fibula + 80.07	±4.02
African female	3.08 * humerus + 64.47	±4.25
	3.67 * radius + 71.79	±4.59
	3.31 * ulna + 75.38	±4.83
	2.28 * femur + 59.76	±3.41
	2.45 * tibia + 72.65	±3.70
	2.49 * fibula + 70.90	±3.80
Asian male	2.68 * humerus + 83.19	±4.16
	3.54 * radius + 82.00	±4.60
	3.48 * ulna + 77.45	±4.66
	2.15 * femur + 72.57	±3.80
	2.39 * tibia + 81.45	±3.27
	2.40 * fibula + 80.56	±3.24
Asian female[a]	3.22 * humerus + 61.32	±4.35
	2.38 * femur + 56.93	±3.57

Source: Forensic Anthropology Training Manual, Karen Ramey Burns, 2007, p.208.
[a] Not all data available for Asian female.

19.6 Follow-Up Questions

1. List possible sources of error in this activity.

2. Why would making a racial determination based on physical traits in the human skeleton be difficult and possibly less reliable than determining other human characteristics?

3. The lengths of the bones in the scenario were determined. Humerus = 32 cm, femur = 46 cm. Determine the maximum and minimum height of the person in feet and inches.

19.7 Forensic Anthropology Worksheet

Data Table

Name of Student	Length of Bone (cm)	Height Range (cm)	Height Range (in)	Average Calculated Height	Measured Height	Percent Difference
	Humerus					
	Ulna					
	Femur					
	Tibia					
	Humerus					
	Ulna					
	Femur					
	Tibia					

Lab

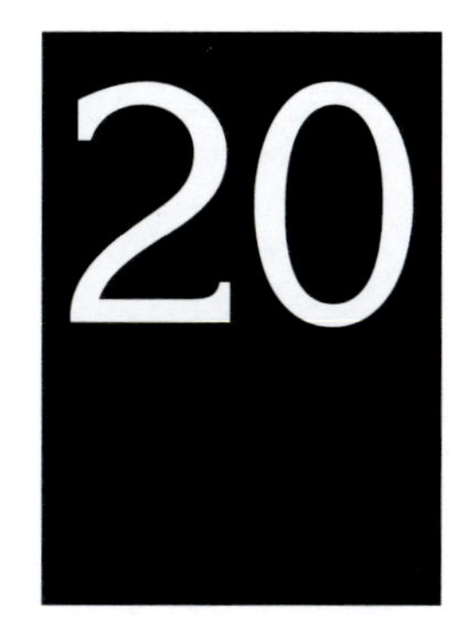

Basic Bloodstain Pattern Analysis

20.1 Background

One technique used by crime scene investigators is the analysis of stains that has been shed at a scene. Bloodstain pattern analysis is a powerful forensic tool used in crime scene investigations. If the investigator understands the dynamics of an altercation, how blood behaves when it exits the body, and how it reacts when it contacts a surface, then an attempt can be made to understand what happened and to determine if a crime occurred. The trained forensic scientist looks at the patterns made by bloodshed and tries to determine what did and/or did not happen. Interpreting the bloodstain patterns involves physical measurement of blood droplets, pattern recognition using known photographs or experiments, the use of trigonometry, and knowledge of the physics of motion. Together with other types of evidence from the crime scene (such as fingerprints, toolmark and footprint impressions, DNA evidence, and chemical analysis), the forensic investigator pieces together the puzzle recreating a logical sequence of events, supported by crime scene evidence. Collecting and documenting the evidence correctly is another skill just as important as interpreting evidence. Bloodstains cannot always be carried back to the lab, so care in documenting the scene is of utmost importance. Photographs and detailed sketches drawn to scale are invaluable tools that help piece together the puzzle.

These activities will introduce you to some of the basics of bloodstain pattern analysis. Part one will ask you to produce typical bloodstain patterns made by vertically dripped blood. Part two analyzes angled blood drops commonly found at a crime scene where blood has been shed. You should sketch and document the patterns produced in each activity. Photographs are recommended so that a visual record can be kept. If this is not possible, attention to detail in sketches is imperative.

Mathematical skills needed for the activities involve taking metric measurements, using a caliper to measure small distances, and knowledge of basic trigonometry.

20.2 Pre-Laboratory Questions

1. Blood is sometimes called a *projectile*. What does that mean? What influences the motion of this projectile?

2. What function does bloodstain pattern analysis play in crime scene investigation?

20.3 Scenario

Police were called to a residence in a quiet suburban neighborhood. The residents stated that they had not seen the family who live there in the last 24 hours and usually they were very active, in and out of the home many times a day. When entering the home, investigators found the four family members deceased in various rooms. There were many circular bloodstains, as well as oval stains, on the walls and floors. Some larger patterns were also noted. A bloodstain analyst was called to collect evidence and analyze the patterns in order to interpret what types of criminal activities and confrontations could have occurred.

20.4 Materials

For Part One

- Simulated drip and projected blood in dropper bottle
- Meterstick
- Poster board or cardstock paper or 5 × 8 note card
- Metric calipers or 15 cm ruler calibrated in millimeters
- 10 ml graduated cylinder

For Part Two

- Simulated drip and projected blood
- 5 × 8 note cards (9)
- 15 and 30 cm rulers
- Clipboard
- Protractor
- Tape
- Metric calipers or 15 cm ruler calibrated in millimeters
- Calculator with trigonometric functions

20.5 Bloodstain Pattern Analysis Laboratory Exercises

20.5.1 Part One—Vertical Drip Pattern Recognition

20.5.1.1 Activity A: Single Blood Drop Patterns

1. Hold the blood dropper bottle upside down in a vertical position, so that the dropper end is 15 cm from the target surface (card).
2. Gently squeeze the bottle so that one drop is released from the bottle at a height of 15 cm and lands on the card.
3. Mark this specimen as vertical drop at 15 cm.
4. Reposition the blood dropper so that the tip is 30 cm from the target surface. Repeat steps 2 and 3, labeling for the correct height.
5. Generate the remaining blood drops at the vertical heights listed in Table 20.1 using the same procedure.
6. Allow the blood drops to dry overnight.
7. When blood drops are dry, measure the diameter of the *circular* part of the drop in millimeters. If there are spines or protrusions on the drop, disregard them as part of the measurement. Record your measurements in Table 20.1.
8. Make a sketch of each drop in Table 20.1, making note of any spines or protrusions in the drop.

20.5.1.2 Activity B: Multiple Blood Drop Patterns

1. Using the simulated blood dropper bottle, drip a single drop of blood onto a target surface from a height of 30 cm. Do not move your hand from the 30-cm position. Drop a second drop of blood onto the first. Note any change in the blood drop pattern after dropping the second drop into the first. Record your observations in Table 20.2.
2. Drop a third blood drop from the 30-cm position, so that the drop lands on the stain from the first two drops. Record your observations.
3. Repeat the procedure until you have mixed a total of four blood drops. Record your observations.
4. If possible, save the target surface and allow it to dry overnight. Label the surface *multiple single vertical blood drops.*
5. Obtain a clean target surface. Place approximately 2–3 ml of simulated blood into a 10 ml graduated cylinder. Position the cylinder 30 cm above the target surface, and pour the entire volume of blood all at once onto the target surface. Observe the pattern and record your observations in Table 20.2. Save the stain pattern and allow it to dry overnight. Label the stain *large volume vertical drip* pattern.

20.5.2 Part Two—Angled Blood Drops

20.5.2.1 Activity A: Making the Angled Blood Drops

1. *Assembling the impact angle clipboard apparatus.* Tape one end of the clipboard to the table by making a hinge out of tape. The clipboard should be right at the edge of the table. Hold the protractor to the edge of the clipboard, so that the protractor is perpendicular to the table top and aligned so that the center of the protractor is at the edge of the hinged end of the clipboard (see Figure 20.1).
2. *Select an impact angle to study.* Table 20.3 includes angles from 10° to 90°. You may select other angles to study. Once you have selected an impact angle, indicate the angle on the card and then attach the card to the board (see Figure 20.1). Set the angle of the board using the protractor, so that

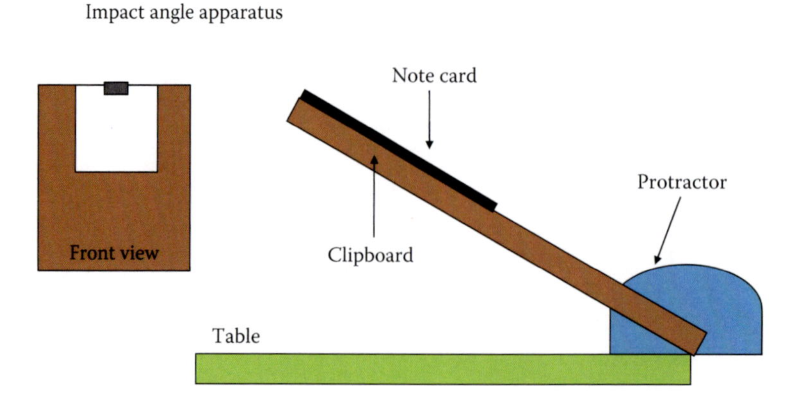

Figure 20.1
Simulating angled blood drops. (Courtesy of Kathy Mirakovits, Forensic Science Educational Consulting, LLC.)

the board will have the correct impact angle. You may need to use a partner for this. To obtain the desired impact angle, set the impact angle board to a protractor setting of 90° minus the desired impact angle. For example, to obtain a 60° impact angle, set the board at a protractor setting of 30°.

3. *Using the simulated drip and projected blood*. Place the dropper at approximately 30 cm above the impact angle board. Allow three drops to fall sequentially onto the card; move your hand so that the drops don't overlap. Using multiple drops will allow you to see the variation that single drops may display when striking the surface. Hold the board at the angle for 8–10 seconds, then lie the clipboard down. Label the card with your name and the impact angle, and then carefully remove the card.
4. Allow the drops to dry and then store the cards in your notebook or in some other safe place.
5. Record your description, including a sketch of the drop in Table 20.3. Note the amount of variation that you see in shape.
6. Repeat Steps 2 through 6 for the remaining impact angles.

20.5.2.2 Activity B: Measuring and Calculating Impact Angles

1. Using the cards you created in Activity A, measure in millimeters the widest part of the bloodstain and the length of the bloodstain. Be careful not to include spines or the tail of the stain. Measurement only includes the length and width of the original blood drop as it hit the surface, not the material that flows after impact (see Figure 20.2).
2. Calculate an average length and width of the three blood drops. Enter your width and length values into Table 20.4.
3. Using the formula in the diagram above, calculate θ, the impact angle.
4. Compare your calculated angles with the known angles of the drop.

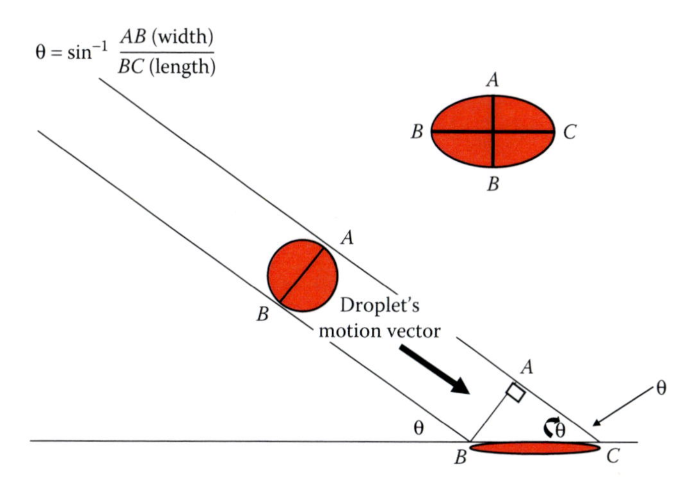

Figure 20.2
Analyzing the motion of a blood drop. (Courtesy of Kathy Mirakovits, Forensic Science Educational Consulting, LLC.)

20.6 Follow-Up Questions

1. How are the blood drops at different heights alike? How do they differ?

2. How do the large volume and dripped volume blood patterns compare?

3. Graph the drop height versus diameter for the blood drops in Part One (Section 20.3.1). What kind of curve is produced?

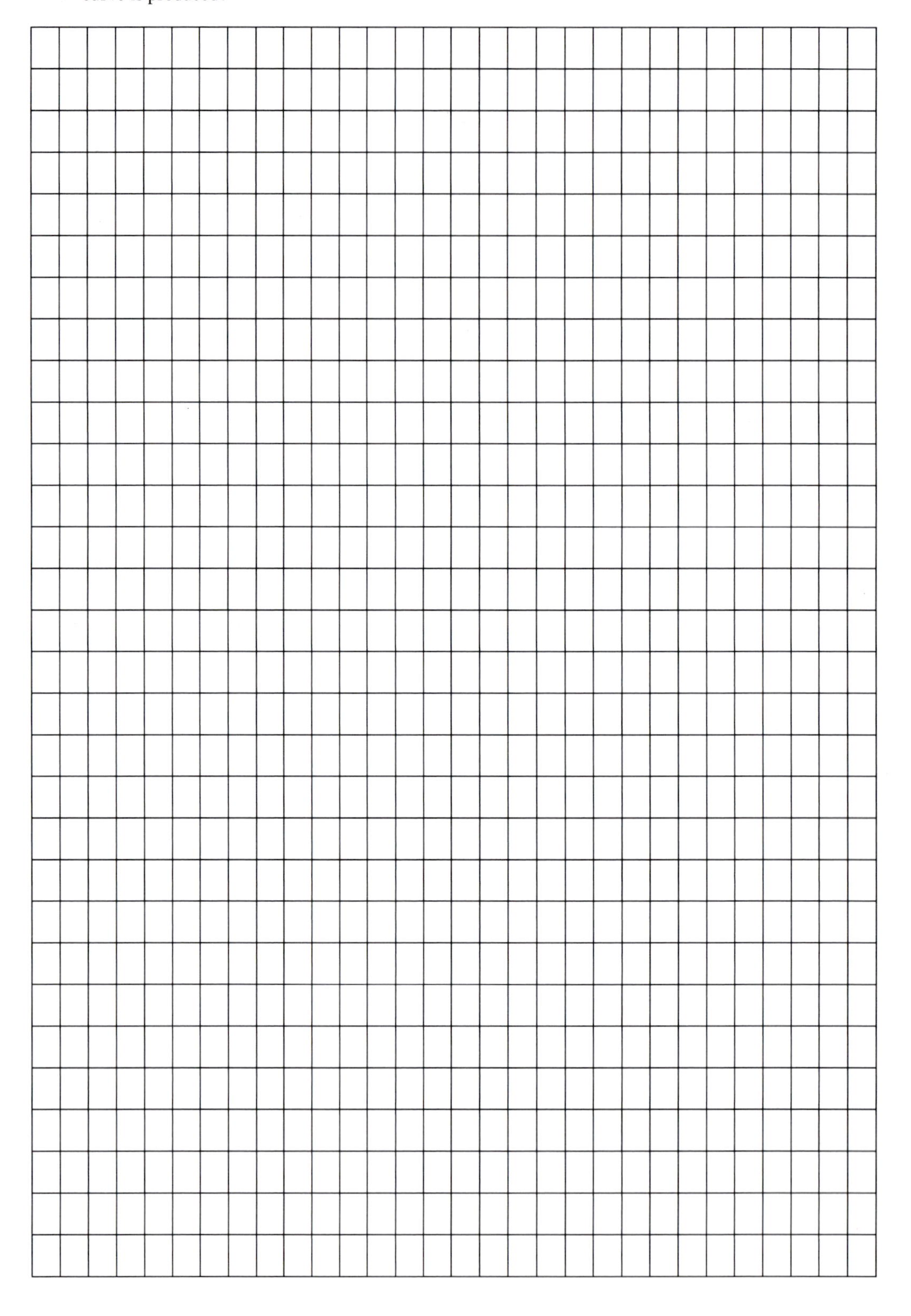

20.7 Basic Bloodstain Pattern Analysis Worksheets

TABLE 20.1
Height, Diameter, and Sketch of Blood Drop

Height of Blood Drop (in centimeters)	Diameter of Blood Drop	Sketch of Blood Drop
15		
30		
45		
60		
75		
100		
150		

TABLE 20.2
Blood Drops and Observations

Blood Drops	Observations—Verbal and Pictorial
One	
Two	
Three	
Four	
Large volume	

TABLE 20.3
Impact Angles

Impact Angle	Description	Sketch
10°		
20°		
30°		
40°		
50°		
60°		
70°		
80°		
90°		

TABLE 20.4
Length and Width of Blood Drops to Calculate the Impact Angle

Impact Angle	Width (in millimeters)	Length (in millimeters)	Calculated Impact Angle
10°			
20°			
30°			
40°			
50°			
60°			
70°			
80°			
90°			

Unit 4

Forensic Chemistry

Lab 21

Inks

21.1 Introduction

Ink analysis is a routine examination conducted by forensic scientists when dealing with questioned documents. While separate inks can look the same on paper without any visible individual characteristics, various methods are used to reveal distinct differences between two seemingly identical lines of ink.

A method used in ink analysis is thin layer chromatography. This is a simple process of physically separating a substance using a physical barrier and a liquid to push the substance through. It is an inexpensive test and can easily distinguish differences in inks by leaving visual cues. The purpose of chromatography is to separate the substance that is being examined based on either polarity of the substance or mass action. There are two components in chromatography, a stationary phase and a mobile phase. The stationary phase acts as a physical barrier for the samples. The mobile phase carries the sample through the stationary phase. This causes the sample to separate into its components over time.

With inks, differences are commonly seen between types and brands of pens. This is because different brands use different colors of dye to make the ink, which results in the color that is observed to be written on the page of paper. Using chromatography, it is possible to separate the inks in order to see if two different samples could have or could not have come from the same type and brand of pen. Not only can the distance each component of an ink sample travels be observed visually, but it can also be determined objectively by calculating the retention factor value, which is the ratio of how far a substance traveled up the stationary phase over the distance the solvent traveled up the stationary phase.

21.2 Pre-Laboratory Questions

1. What are the two phases in chromatography?

2. What is the purpose of chromatography?

3. Would it be expected for two Bic® black pens to have the same components?

21.3 Scenario

A ransom note was found at the location of the missing puppy. There was thought to believe that an employee of animal control was kidnapping all dogs in the neighborhood. A warrant was obtained for the suspect's home and multiple pens were collected. The forensic analyst is asked to determine in any of the pens found in the suspect's home could have written the ransom note found at the crime scene.

21.4 Materials

- Chromatography chamber
- Aluminum foil
- Pencils
- Filter paper
- Variety of pens
- Ruler
- Isopropyl alcohol

21.5 Procedure

1. Ink analysis
   a. Add 5 ml isopropyl alcohol to a beaker and cover it with aluminum foil.
   b. Draw a pencil line 10 mm up from the bottom of the filter paper. Mark evenly spaced dashes along line on the paper.
   c. Take the different pens and make a dot on the line of the filter paper by the dashes.
   d. Allow samples to thoroughly dry.
   e. Make sure that there is 5 mm of isopropyl alcohol in the beaker but make sure that the amount is below the marked line on the filter paper.
   f. Place the filter paper in the beaker, being careful not to touch or cling to the sides of the beaker or splash the solvent onto the paper.
   g. Cover the beaker with aluminum foil and allow the mobile phase to move up the paper.
   h. Once it has moved three-fourths of the way up, remove the paper from the beaker and draw a line with pencil at the mobile phase stopping point.
   i. Allow the filter paper to dry completely.
   j. Record the color and shape of the components of the ink.
   k. Mark the center of the spot or band of each ink sample created on the paper.
   l. Measure the distance between the bottom line and the spot marked. Record this measurement. Do this for each component.
   m. Calculate the retention factor value for each spot.

$$\text{Retention factor} = \frac{\text{Distance traveled by the substance}}{\text{Distance traveled by the solvent}}$$

21.6 Follow-Up Questions

1. Why is handwriting analysis performed before ink analysis chromatography?

2. Looking back at your thin layer chromatography plate, explain how you determined that your ink samples came from the same type of pen or different types of pens.

3. On a developed filter paper, a single spot was observed for specific ink sample. The solvent was allowed to travel 30 mm up the filter paper from the spotting of the ink sample. The ink sample was measured to have traveled 19 mm up the filter paper. What is the calculated retention factor value for this ink sample?

21.7 Inks Worksheet

Draw your thin layer chromatography plate. Include all retention factor values, calculations, measurements, and labels to your diagram.

Lab 22 Illicit Drugs

22.1 Introduction

An illicit drug is one that has been considered illegal to have in possession. These drugs are separated based on federal schedule, effect, and composition. There are five schedules of drugs as controlled substances under the Controlled Substances Act, Schedules I–V, which discriminate drugs based on medical use, potential for abuse, and psychological and physical dependence. The most dangerous illicit drugs, which possess no accepted medical use in the United States, very high potential for abuse, and can cause severe psychological/physiological dependence, are categorized into Schedule I. Drugs can also be categorized by the effect they have when ingested by a person. Classification of drugs can also be performed based on the composition, or how it is physically manufactured. Drugs are divided into three categories: natural, synthetic, and semisynthetic.

The analysis of drugs in general has become highly structured and objective by the forensic scientific community through the establishment of a common analytical scheme. The scheme consists of categorizing the different tests used into different classes based on the discriminating capacity of the test. Class C tests consist of simple screening tests, which can be used to presumptively determine if a drug is present in the sample being analyzed. Class B tests consist of separation tests, often in the form of chromatography, which can lead to the identification of a drug through the separation of the drug of interest from other substances within the sample. Class A tests consist of confirmatory tests, which will result in a positive test for only a specific drug. Class A tests are based on complex compound classifications.

During this lab exercise, both Class B and Class C presumptive tests will be performed on a variety of drugs.

22.2 Pre-Laboratory Questions

1. How many schedules of drugs are there and how is each one determined?

2. What are the three classes of analysis for drugs? Give the difference between each class.

3. What are the three types of drugs and how are they different from each other?

22.3 Scenario

Police officers received information about a possible drug laboratory going in one of the old abandoned warehouses downtown. Upon the officers' arrival, they found several pieces of equipment and laboratory supplies that could be used to illegally produce drugs. Several small plastic bags of white powder were also found in the laboratory. The items that were found in the warehouse were sent to the drug analyst at the crime lab and need to be analyzed for the presence of any types of illicit drugs. The items were then examined using presumptive and confirmatory tests to check and confirm the presence of any drugs.

22.4 Materials

- Spot plates
- Reagents
- Excedrin (copyright symbol) tablets
- Chromatography chamber and supplies

22.5 Laboratory Procedures

22.5.1 Basic Laboratory Procedure

1. Spot tests/color tests
   a. Place a small amount (pinch) of each white powder in a well of a spot plate.
   b. Make sure to label what is in each well.
   c. Add a droplet of reagent to each powder.
   d. Record the reactions observed.
2. Chromatography
   a. Crush an Excedrin tablet and dissolve with water.
   b. Add 5 ml isopropyl alcohol to a beaker and cover it with aluminum foil.
   c. Draw a pencil line 10 mm up from the bottom of the filter paper. Mark evenly spaced dashes along line on the paper.
   d. Using a capillary tube, draw up the dissolved Excedrin, then spot the capillary tube onto the pencil line on the filter paper; you may want to make three spots.
   e. Make sure that there is 5 mm of isopropyl alcohol in the beaker but make sure that the amount is below the marked line on the filter paper.
   f. Place the filter paper in the beaker, being careful not to touch or cling to the sides of the beaker or splash the solvent onto the paper.
   g. Cover the beaker with aluminum foil and allow the mobile phase to move up the paper.
   h. Once it has moved three-fourths of the way up, remove the paper from the beaker and draw a line with pencil at the mobile phase stopping point.
   i. Allow the filter paper to dry completely.
   j. Spray the filter paper with the iodoplatinate spray.
   k. Allow the filter paper to dry completely; a hair dryer on the lowest setting can be used to speed up the process.

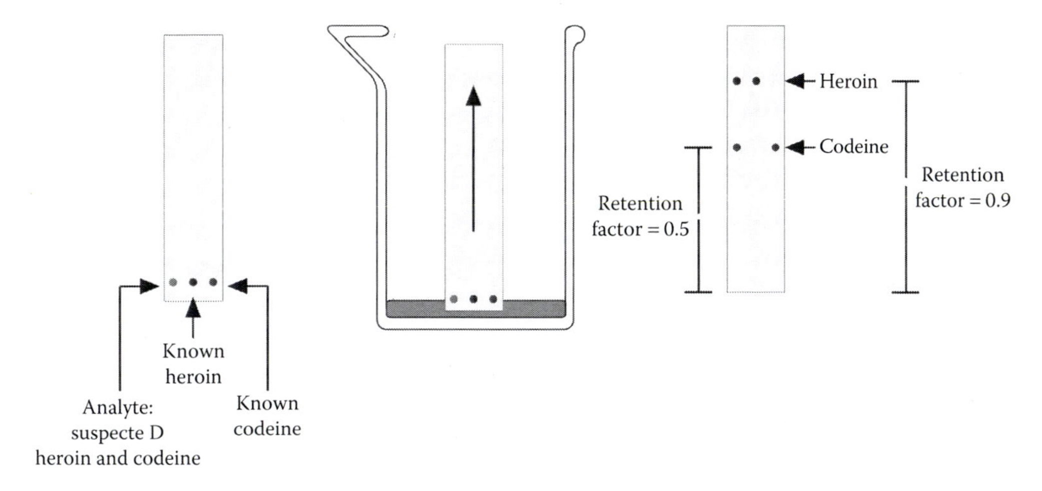

Figure 22.1
Chromatography experiment.

l. Mark the center of the spot or band of each component in the Excedrin created on the paper.
m. Measure the distance between the bottom line and the spot marked. Record this measurement. Do this for each component.
n. Calculate the retention factor value for each spot.

$$\text{Retention factor} = \frac{\text{Distance traveled by the substance}}{\text{Distance traveled by the solvent}}$$

See Figure 22.1 for demonstration of preparing the chromatography experiment.

22.5.2 Advanced Laboratory Procedure

1. Microcrystal tests
   a. Add a pinch of unknown white powder to the center of a microscope slide.
   b. Add a drop of 3M hydrochloric acid to the white powder.
   c. Add a drop a gold chloride reagent on the microscope slide next to the dissolved unknown white powder.
   d. Using a toothpick, drag the gold chloride reagent drop into the dissolved white powder drop.
   e. After two minutes, place the microscope slide under the microscope at 100× magnification.
   f. Record any observations seen (shape, size, abundance, etc.), including drawings of the crystals on worksheet.
   g. Repeat steps a through f with other unknown white powders.
   h. Compare the drawings of the unknown white powders to the known images of microcrystals from procaine, lidocaine, and procainamide.
   i. Identify each unknown on the worksheet.

22.6 Follow-Up Questions

22.6.1 Basic Follow-Up Questions

1. How many components were discovered in the Excedrin? How did you determine this?

2. What are the components of Excedrin? If your analysis was different, why?

3. Are chromatography and reaction test confirmatory tests for drugs, why or why not?

22.6.2 Advanced Follow-Up Questions

1. What observable characteristics where there between the different drugs for the microcrystal test?

22.7 Illicit Drugs Worksheet

Draw your thin layer chromatography plate. Include all retention factor values, calculations, measurements, and labels to your diagram.		
Sample		
Color		
Transparency?		
Crystal description or characteristics		

Lab

White Powder Testing

23.1 Introduction

There are many compounds that are white and powdered. Some over-the-counter drug compounds are white and come in powdered form or can be ground into powdered form. This activity analyzes basic white powders, some food-based and others from over-the-counter drugs.

23.2 Scenario

A call came in to the police from one Mr. Smith, who claimed that his home was broken into while he was cooking in the kitchen and he was threatened and beaten. When investigators arrived, they found white powders of various textures all over the kitchen. Mr. Smith claimed that he was baking for a bake sale and that is why there was so much powder everywhere. The neighborhood was one that had been known for drug activity; therefore, many samples of the powders were collected and are in the lab for analysis. No suspects have been noted yet in this case.

23.3 Materials

- Spot plates
- Microspatulas
- pH Hydrion paper and color chart
- Toothpicks
- Gloves
- Safety goggles

White Powders to Test	Reagents
Acetaminophen (Tylenol®)	2% ferric chloride solution
Alka-Seltzer	2% cobalt thiocyanate solution
Acetylsalicylic acid (aspirin)	1% HCl solution
Benadryl	Lugol's iodine
Baking soda	Distilled water
Cornstarch	
Ibuprofen (Motrin®)	
Sugar	
Vitamin C	

- Crime scene evidence to identify given by the instructor

23.4 Procedure

Safety note: Wear goggles and gloves when conducting this activity.

1. Use a well plate to test three powders at a time.
2. Place a *small amount* (about the size of a grain of rice) of one powder into a row of five wells. See example below:

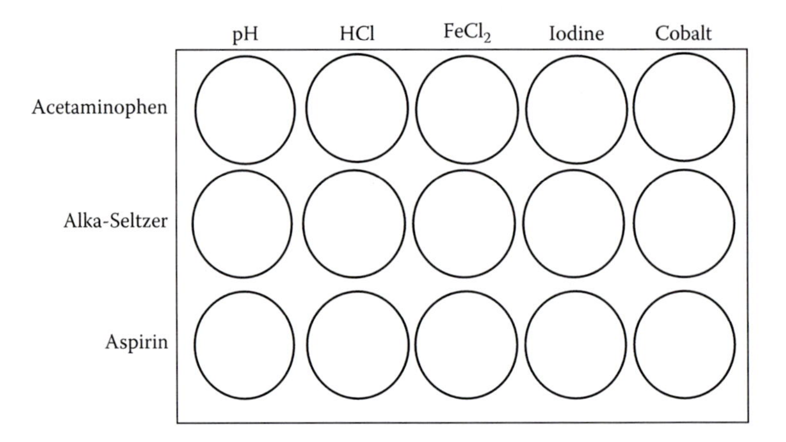

3. Test in the following order:
   a. *Well 1:* Look at the appearance—is the substance *grainy or powdery* and *pure white or has a color*? Record the observations in the chart.
   b. *Well 1:* Add one or two drops of water. Stir with a toothpick. Does the substance react? Does it dissolve readily and float? Record the observations in the chart.
   c. *Well 1:* Dip a piece of pH paper in the water/powder mix. Remove the paper and record the color of the paper and the resulting pH value in the chart.
   d. *Well 2:* Add one or two drops of hydrochloric acid to the powder. Observe for any reaction, stir, and record your results.
   e. *Well 3:* Add one or two drops of ferric chloride to the substance. Observe and record the color and clarity (clear, cloudy, and grainy) of the mixture.
   f. *Well 4:* Add one or two drops of Lugol's iodine to the substance. Observe and record the color and clarity of the mixture.
   g. *Well 5:* Add one or two drops of cobalt thiocyanate to the substance. Observe and record the color and clarity of the mixture.
   h. Rinse and dry your well plate and repeat with the next three powders. Repeat for the last three powders.

23.5 Follow-Up Questions

1. Which of the powders were acidic? Which were basic?

2. Construct a chart of one or two test results that were *unique* to that powder; a result that would single out that powder from the rest.

3. Report out on the crime scene evidence.
   How do the results affect the direction of the investigation?
   Be specific with evidence to reinforce your statements.

23.6 White Powder Testing Worksheet

Unknown Powder Lab—Over-the-Counter Drugs Table

Name of the Powder	Appearance	Reaction with H_2O	pH	Reaction with HCl	Reaction with Ferric Chloride	Reaction with Iodine	Reaction with Cobalt Thiocyanate
Acetaminophen (tylenol)							
Alka-Seltzer							
Acetylsalicylic acid (aspirin)							
Benadryl							
Baking soda							
Cornstarch							
Ibuprofen (motrin)							
Powdered sugar							
Vitamin C							

Lab 24

Calculating Blood Alcohol Concentration
The Widmark Formula

24.1 Introduction

E.M.P. Widmark, a Swedish researcher in pharmacokinetics, developed this equation for calculating a person's blood alcohol concentration (BAC) in the early twentieth century. It is still widely used today, often in court when an alcohol-related crime is being brought to trial. The formula takes into account many factors. First of all, the percentage of alcohol in the drink consumed is a major factor, as liquor shots are higher percent alcohol than a typical beer. The gender and weight of the person is another large factor. Males are affected less than females by alcohol content, since females have more fat cells and less water content than males. Fat does not absorb alcohol and therefore the alcohol gets absorbed into the bloodstream of the female more readily. If the female has less percentage of water in her body compared to the male, this also lends itself to a higher concentration of alcohol from drinking the same amount of drinks. Women are generally smaller in size than men, and this factor causes females to have a higher concentration of blood alcohol. All of these factors are part of the Widmark formula.

Additionally, the time since drinking commenced will affect the BAC. If the same amount of alcohol is consumed in a longer time interval, the individual's body will have time to process the alcohol and clear it from the blood and body. As a result, the BAC will be lower.

24.1.1 Basic Widmark Formula

- A is the alcohol consumed in ounces (oz.)
- D is the percentage of alcohol written as a decimal (e.g., 6% = 0.06 and 40% = 0.40)
- Wt is the weight of person in pounds (lb.)

$$\text{Male formula:} \quad \frac{(A)(D)(5.14)}{(\text{Wt})(0.73)}$$

$$\text{Female formula:} \quad \frac{(A)(D)(5.14)}{(\text{Wt})(0.66)}$$

- *Beer*: 12 oz. serving is average at the rate of 6% alcohol
- *Wine*: 5 oz. serving is average at the rate of 12% alcohol
- *Hard liquor (shot)*: 1.25 oz. is average at the rate of 40% alcohol (80 proof)

24.1.1.1 Time Factor

When alcohol is consumed over a long period of time, the following additional calculation is done:

1. Estimate the hours since the drinking *commenced.*
2. Multiply the hours × 0.015.
3. Subtract this value from the BAC calculation.

24.1.2 Widmark Formula with Time Consideration

$$\text{Male formula: } \frac{(A)(D)(5.14)}{(\text{Wt})(0.73)} - (0.015 \times \text{hours})$$

$$\text{Female formula: } \frac{(A)(D)(5.14)}{(\text{Wt})(0.66)} - (0.015 \times \text{hours})$$

24.1.2.1 Example Problems

A 185-pound *male* consumes one beer.

$$\frac{(12\text{oz.})(0.06)(5.14)}{(185)(0.73)} = \frac{3.7008}{135.05} = 0.03\,\text{BAC}$$

A 120-pound *female* drinks one beer.

$$\frac{(12\text{oz.})(0.06)(5.14)}{(120)(0.66)} = \frac{3.7008}{79.2} = 0.05\,\text{BAC}$$

24.1.3 Problems: Show Your Work

1. A 150-pound man has two beers. Determine his BAC.

   BAC =

2. A 112-pound woman has three glasses of wine. Determine her BAC.

   BAC =

3. The same 112-pound woman in Problem 2 has three beers. Determine her BAC.

   BAC =

4. The same 112-pound woman in Problem 2 has three rum and cokes. Determine her BAC.

   BAC =

5. How many ounces of beer would a 200-pound man need to consume to have the same BAC as the woman in Problem 3?

   Ounces =

   How many beers would this be?

6. A 150-pound male consumed a six-pack of beer over 2 hours and 15 minutes. Determine the BAC for this person.

   BAC =

7. A 110-pound female consumed two glasses of wine in a two-hour period of time. Determine her BAC.

   BAC =

8. A 140-pound female has three mixed drinks one right after the other.

   a. What is her BAC after consuming the three drinks?

      BAC =

   b. What is her BAC three hours after finishing the last drink?

      BAC =

9. A police officer is investigating an auto accident. The officer looks at the evidence file and finds a tab from a local bar in his pocket. The bar tab was started at 9:15 PM and it shows that the 185-pound deceased male had five beers. The accident occurred at 2:30 AM. At the time of the accident, what was the BAC of the person?

   BAC =

10. A 125-pound woman is pulled over by a police officer. The officer gives her the field sobriety tests, and then the officer gives her a BAC test. The test indicates a BAC of 0.085. The woman claimed she drank only two glasses of wine in the past four hours. If she truthfully drank wine and it was within the last four hours, then how many glasses of wine did she really have?

    Glasses of wine =

Lab 25

Fibers

25.1 Introduction

A fiber is defined as a material that has a length that is at least 100 times its diameter. Many objects that are encountered and thought of as fibers are *textile fibers*. A textile fiber is a unit of matter, either natural or manufactured, that forms the basic element of fabrics and other textiles. Overall, there are many different kinds of fibers, which can be categorized in different ways. First, fibers can be placed into one of two basic types, natural or manufactured. Natural fibers are materials that come from animals, plants, or minerals. Manufactured fibers are fibers that have been made or derived by a process in which at any point the material is not a fiber.

Fibers can also be placed into one of four classes, which can further distinguish the type of fiber further than just saying the fiber is natural or manufactured. Fibers can be classified as protein, cellulose, mineral, or synthetic fibers. Protein fibers are those that are made up of polymers, or long repeating chains of connecting strands of amino acids. An example of this is hair, which is primarily made up of the protein known as *keratin*. Essentially, all hair fibers, such as wool, are protein fibers. Cellulose fibers are those that contain polymers of carbohydrates, such as cotton or hemp. Mineral fibers are composed of components of rocks, specifically silica, which can be found in rocks such as quartz. Mineral wool and asbestos are two common examples of mineral fibers. Synthetic fibers are those that are produced using small organic molecules. A common example of a synthetic fiber is polyester.

Although microscopic examinations are preferred for examinations of fibers due to the nondestructive nature of the assessment, another test that can be conducted on fibers is the burn test. Although destructive, burn tests can be used to distinguish between fibers of different chemical compositions based on differences in flame color, burn rate, odor, and residue from the burned material.

Since only class characteristics are observed for fibers, fibers cannot be individualized to a specific textile, but only associated to a textile. This means that fiber analysis can only conclude if a fiber could have come from another textile sample; it cannot determine if a specific fiber came from a specific textile sample with the exclusion of every other textile sample of the same type.

25.2 Pre-Laboratory Questions

1. Give the definition of a fiber and an example of a fiber.

2. What is the difference between a natural fiber and a manufactured fiber? Give an example of each.

3. What are the four classes of fibers?

25.3 Scenario

A house outside of the city was under renovation and was reported of being broken into. The crime scene investigators noticed that throughout the house there were several unique types of fibers. When detectives interviewed the neighbor to collect additional information about a possible burglary, it was noted that the fibers collected at the crime scene were found to look similar to the sweater the neighbor was wearing. Each fiber was collected and stored in its own envelope and sent to the crime lab for analysis.

25.4 Materials

- Metal forceps
- Glass watch plate
- Flame source; candle or alcohol lamp
- Natural and synthetic fibers

25.5 Procedure

1. Burn test
    a. Using tweezers to hold the fiber slowly approach the flame. Note the color of the flame and the behavior of the fiber.
    b. Record the characteristics of the fibers on the chart.

25.6 Follow-Up Questions

1. What would be your method of analysis if you were given a fiber to examine?

2. Can you be sure a fiber came from a specific sample? Why?

3. Describe the results of one of the burn tests that you performed on a fiber.

25.7 Fibers Worksheet

SAMPLE	
Macroscopic Characteristics	
Color	
Length	
Burn Test	
Odor	
Flame color	
Smoke presence	
Burn rate	

Lab 26 Polymers

26.1 Introduction

Polymers are composed of long chains of repeating molecules. There are many types of polymer evidence such as plastics, paints, fibers, and other materials.

Automobile paints are a common type of trace evidence seen in a forensic science laboratory. Car paints are sprayed on during the manufacturing process and have up to three separate layers, all with different composition and purpose. The first layer is a resin that binds to the metal frame of the vehicle, which also aids in adhering to the second layer. The second layer contains the color of the car. And the last layer is a clear coat with an ultraviolet protectant, which keeps light from fading the color of the car.

Solubility testing can be an important test done on paint chips to determine layer composition. Paint layers will react differently to a variety of solvents, which can discriminate between paints chips with different pigments, binder composition, and other additives. Paints are manufactured with different components and therefore will react differently and can be compared to each other. Common reactions include the paint being soluble in the solvent, color bleeding for the paint, the chip curling, swelling, softening, sinking, or floating. The paint may also start to dissolve or bubble in the solvent. These results can lead to identifying the types of acrylics, enamels, or pigments in the paint.

26.2 Pre-Laboratory Questions

1. What is a polymer?

2. How many layers are in automobile paints? Which layer has the color?

3. Name a reaction observed in solubility testing?

26.3 Scenario

A person was hit by a vehicle last night. The victims had fragments of paint chips on their clothing and were collected by crime scene investigators. An eyewitness saw a car strike the person and then crash into a telephone poll at the end of the road. Forensic scientists are asked to compare the paint chips found on the victim with the paint from the crashed vehicle.

26.4 Materials

- Paint chips
- Metal forceps
- Spot plate
- Various solubility solutions

26.5 Procedure

1. Solubility test
   a. Place different paint chip into a ceramic welled plate.
   b. Add a drop of solvent into each well.
   c. Watch and record the reactions.

26.6 Follow-Up Questions

1. What were the different reactions seen between the different paint chips?

2. Why would you expect the reactions to be different between the paint chips?

26.7 Polymers Worksheet

SAMPLE	
Macroscopic Characteristics	
Color	
Length	
Solubility Test Results	
Float	
Color Change	
Curling	
Dissolvable	
Other	

Lab 27

Fire Debris

27.1 Introduction

Essentially, all fires result from the chemical reaction of a fuel and oxygen, known as *combustion*. In combustion, the fuel combines with oxygen to form carbon dioxide, water, and energy. In explosions, the fuel and oxygen are confined closely together, being physically or chemically mixed, causing a very rapid combustion to occur. With a fire, the fuel and oxygen are physically and chemically separated, resulting in slow combustion, since it takes longer for the fuel and oxygen molecules to interact in order to cause the combustion reaction. Therefore, although both are consequences of combustion, the difference between an explosion and fire is the speed in which the combustion reaction occurs, or the detonation velocity.

Arson investigators are responsible for determining whether or not a fire was intentionally set as a crime or caused by natural or accidental means. Natural fires are fires that occur by nature, such as by a lightning strike. Accidental fires are fires that are set without malicious intent, such as dropping a lit cigarette onto a flammable couch after falling asleep. Arson is when a fire is intentionally set with criminal intent, such as burning a house down to collect insurance money. When investigating an arson fire, indicators such as accelerants and multiple points of origin are often found. During an investigation, indicators such as burn patterns are looked for, which leave clues about the fire. Burn patterns can show both the point of origin of a fire, as well as the nature of the fire. To determine the point of origin, or even multiple points of origin, clues such as low burning, V-patterns of the smoke, the charring of wood, the spalling of plaster or concrete, material distortion, and soot and smoke staining are looked for and documented. These clues can lead an investigator to the source of the fire. From here, indicators of arson may include the presence of an accelerant, the elimination of natural or accidental causes of a fire, fire trails, or even multiple points of origin. All of these clues put together can tell the investigator a story about the nature of the fire and help explain how the fire started.

To identify what accelerants may have been used to start a fire, an examination of fire debris is conducted. Before the accelerant can be analyzed, it must be separated from the surface it was placed onto through extraction. The extract is analyzed using chromatograph, which produces a graph of peaks that shows the different components in the accelerant. Different chemicals produce different combinations of peaks from which the pattern of these peaks can be used to identify what accelerant was used in a fire.

27.2 Pre-Laboratory Questions

1. Combustion is the chemical reaction between what two things?

2. List four burn patterns that help show where the point of origin of a fire is located.

3. List four burn patterns that help show indications of an arson fire.

27.3 Scenario

Last night the fire department was called out to the pizza place on the northwest side of town because the kitchen caught on fire. Fire scene investigator, Levi Corduroy, was then called to the scene after the fire was extinguished to examine the scene to determine if the fire was intentionally caused because the business was in financial trouble or was it accidental.

27.4 Materials

- Pictures of fires scenes
- Chromatographs of various types of ignitable liquids

27.5 Procedure

1. Fire residue analysis
   a. Examine the pictures that are provided.
   b. Determine the point(s) of origin of the fire. Look for V-patterns on walls, charring of wood, spalling of plaster or concrete, material distortions, and soot or smoke staining.
   c. Determine and record if the fire was arson. Explain your reasoning including observations such as multiple points of origin, trails of fire, presence of natural or accidental causes of fires, and so on.
2. Instrumental analysis
   a. Examine the graphs from a gas chromatograph (GC) of known accelerant samples. Use pattern recognition to compare these samples to the provided unknown GC graphs.
   b. Using pattern recognition, determine what accelerant each unknown graph can be identified as.

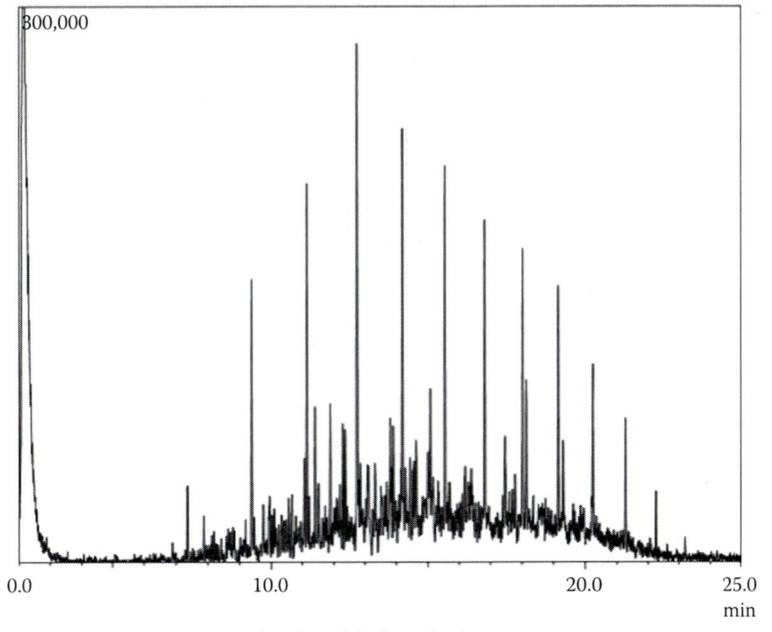

Diesel fuel standard

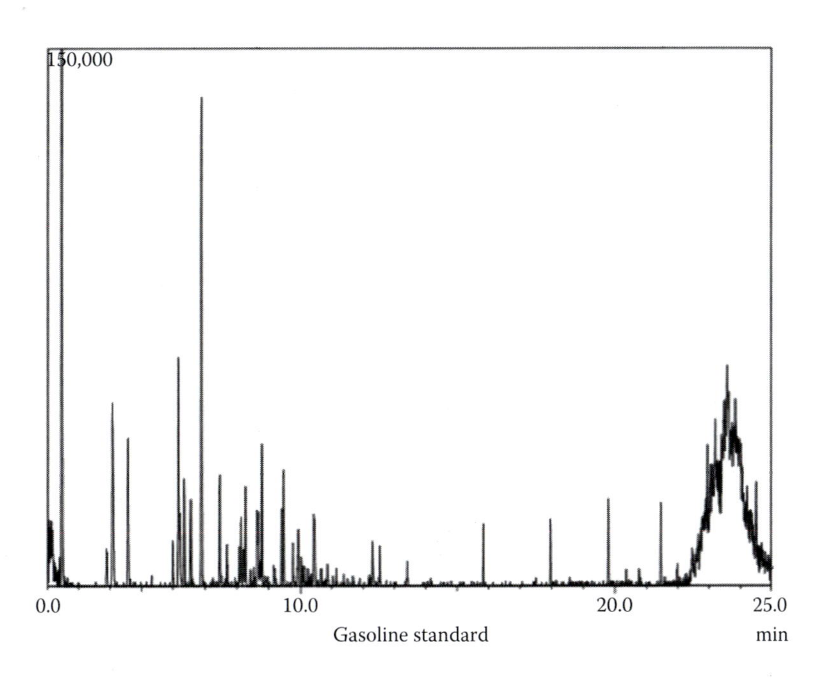

Gasoline standard

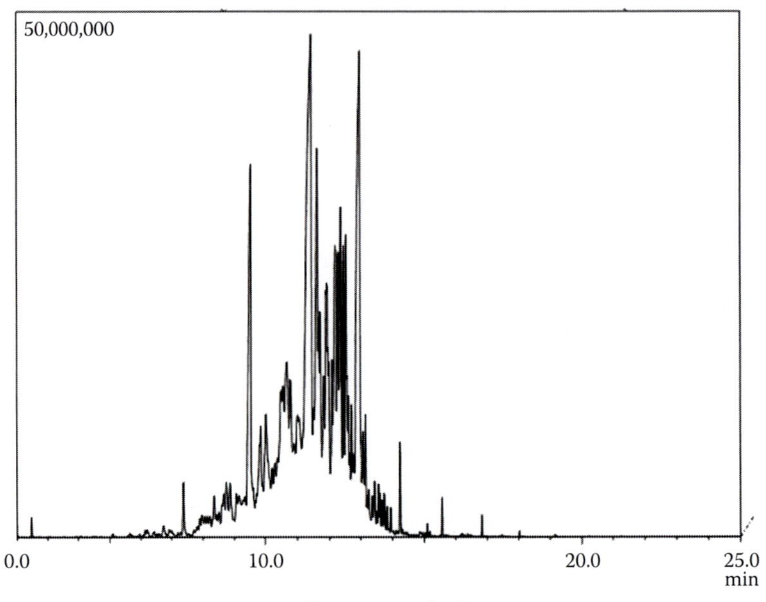

Kerosene standard

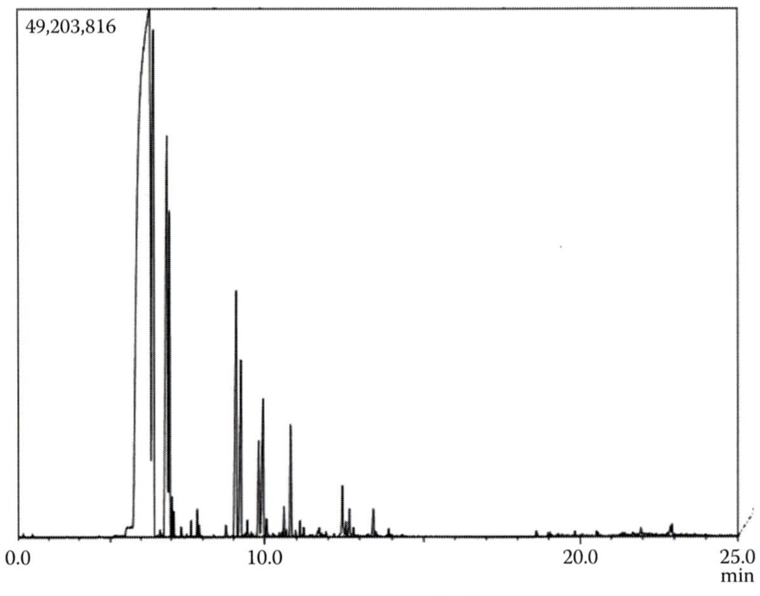

Turpentine standard

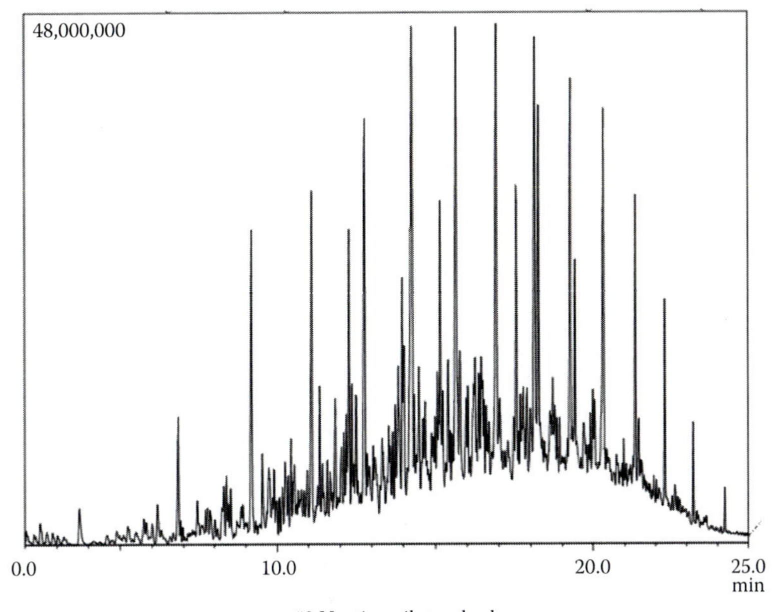

#2 Heating oil standard

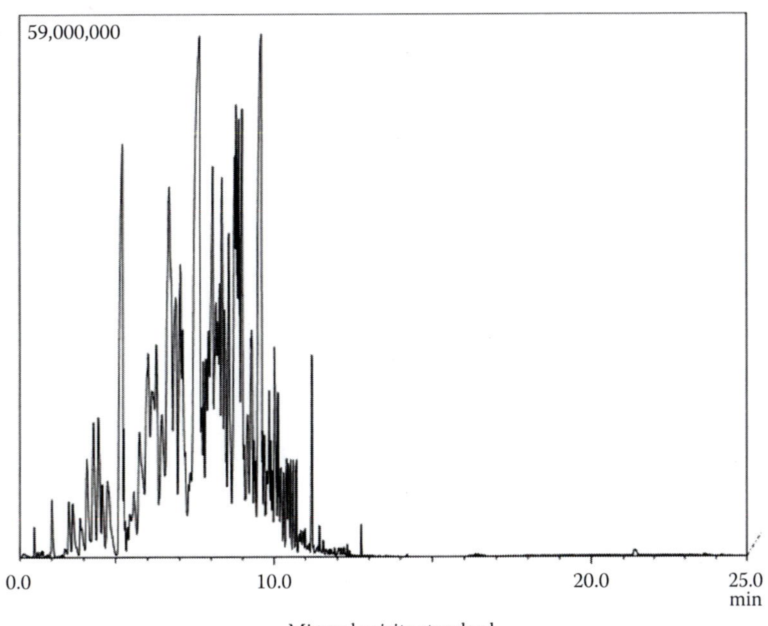

Mineral spirits standard

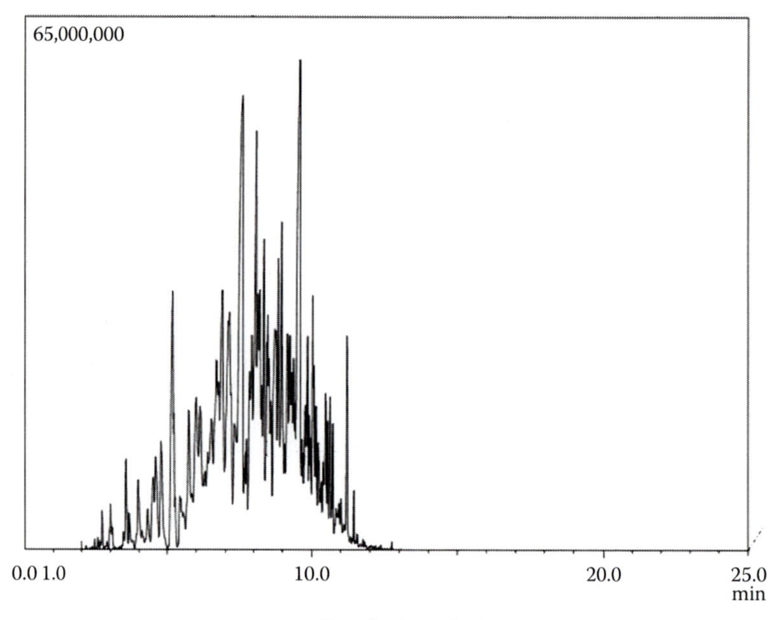

Lighter fluid standard

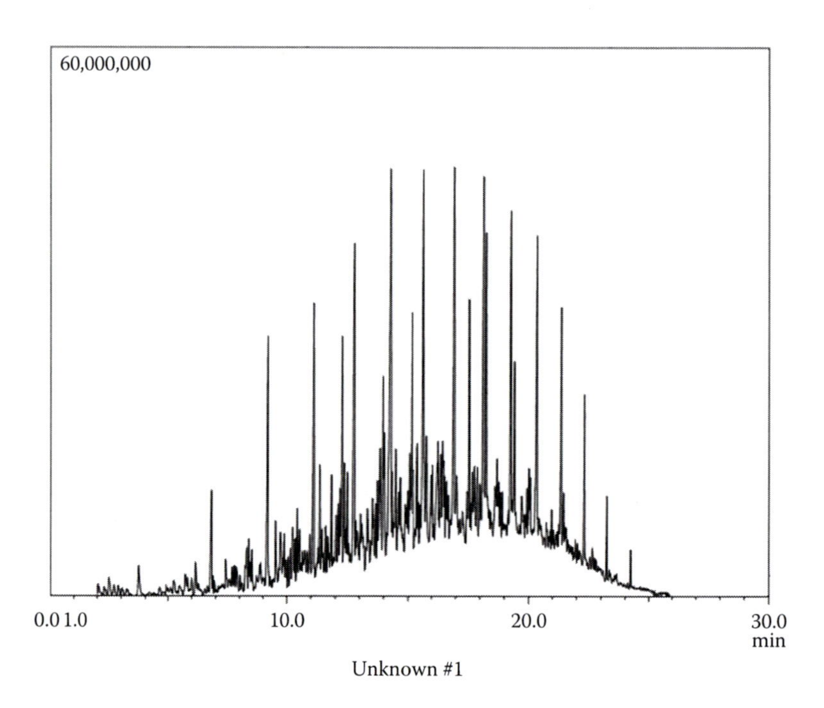

Unknown #1

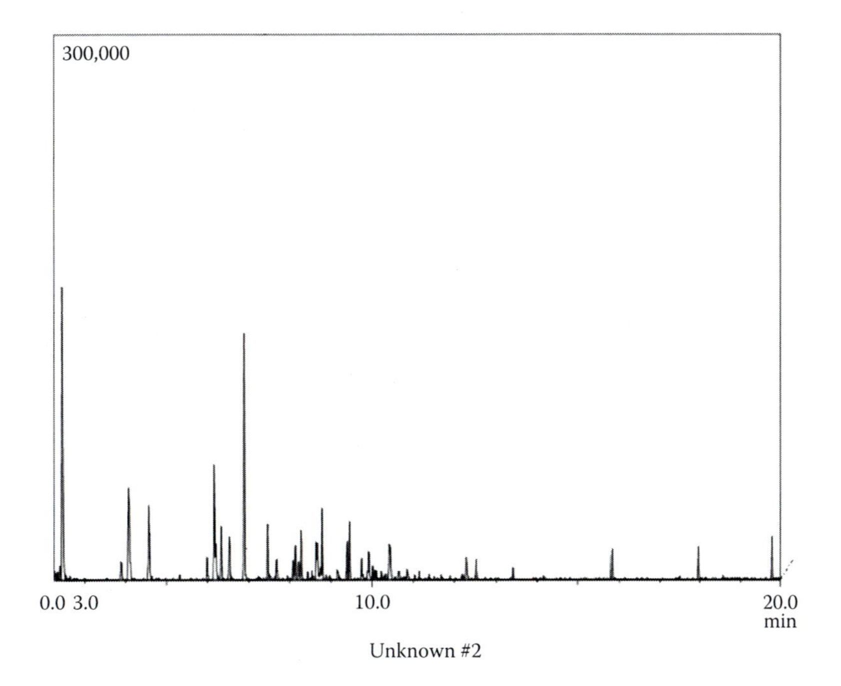

Unknown #2

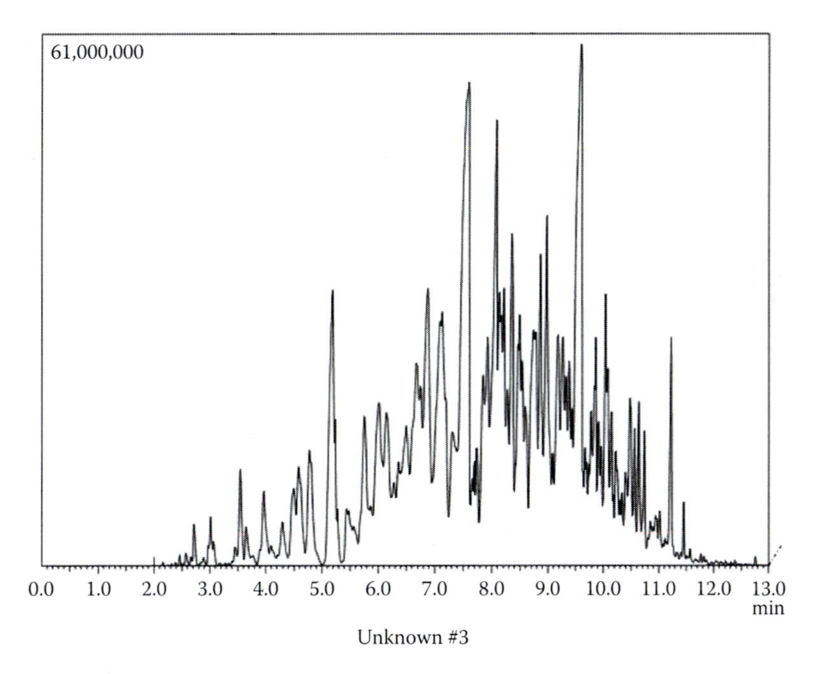

Unknown #3

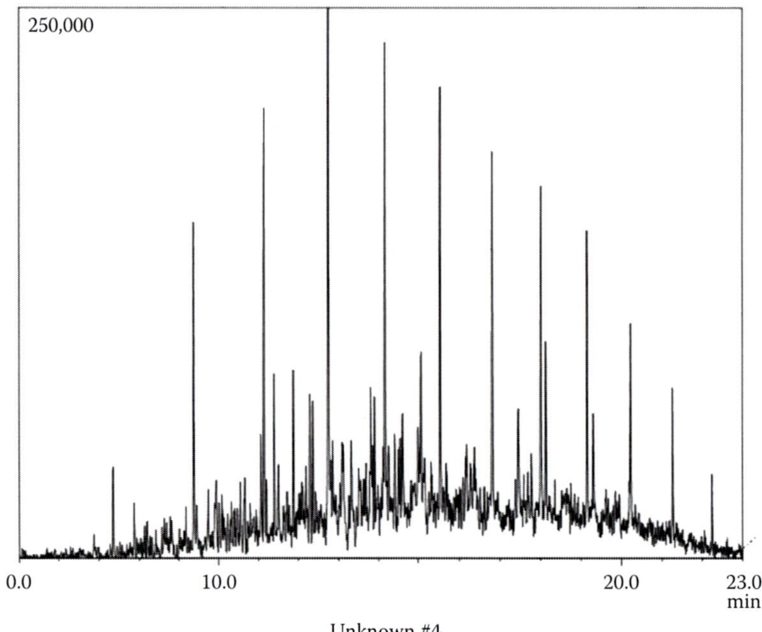

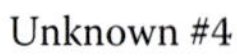
Unknown #4

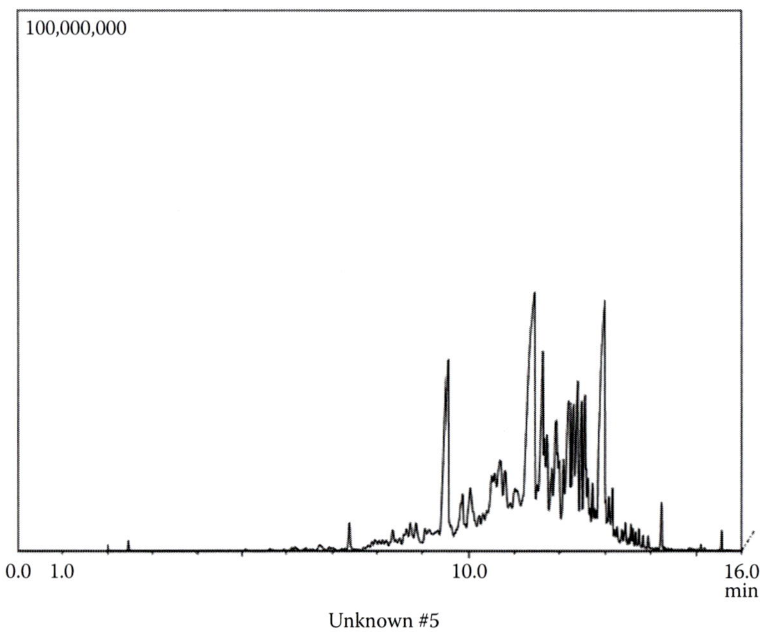

Unknown #5

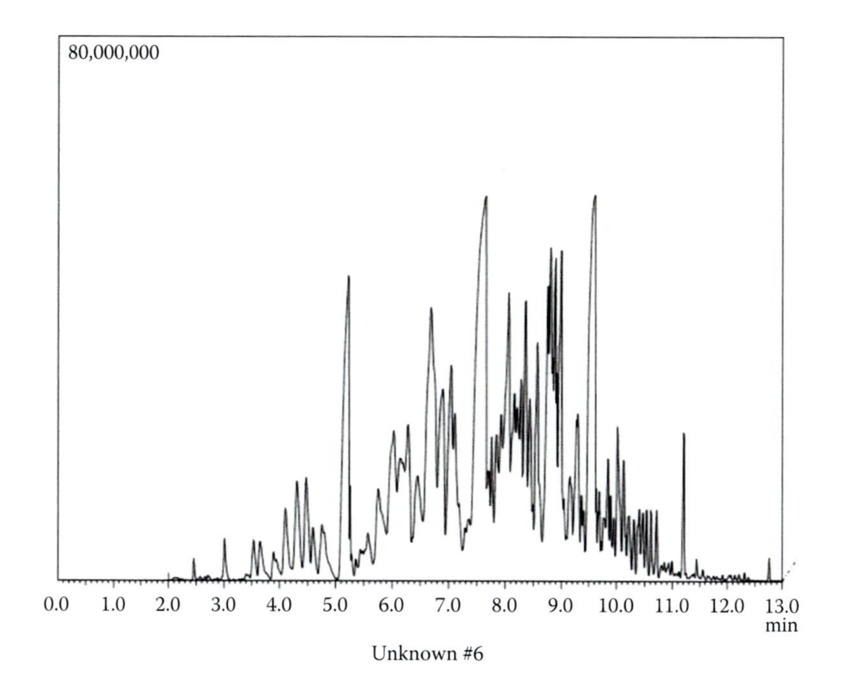

Unknown #6

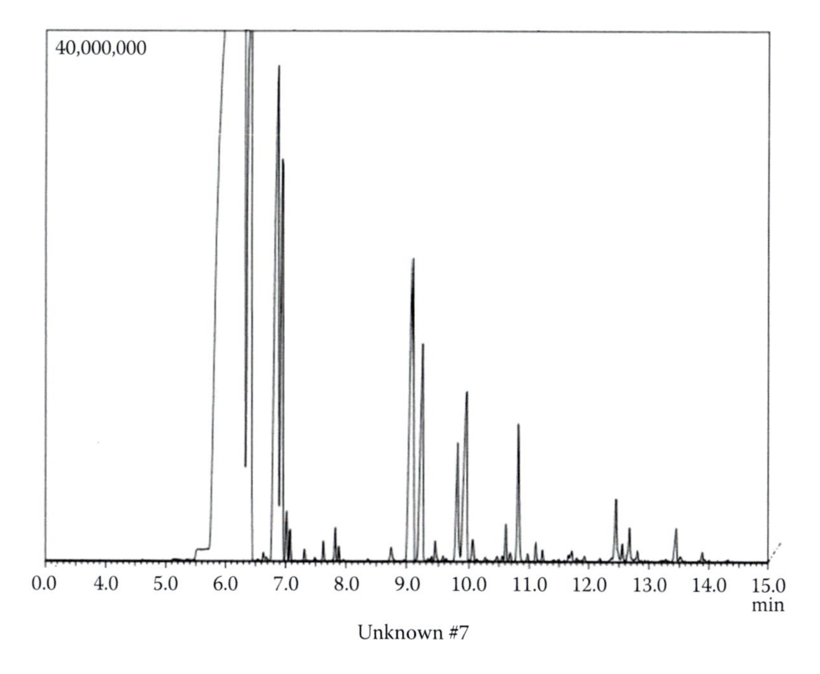

Unknown #7

Figure 27.1
Photographs (a) A fire in a kitchen. Note the V pattern of burning on the wall of the left side. (b) A fire trail in a mobile home.

Photographs (Figure 27.1a and b) can be used for examples.

27.6 Follow-Up Questions

1. In your crime scene pictures, what indicators did you find that showed the possibility of a natural/accidental fire or an intentionally set fire?

2. How did you identify what accelerant was found at the crime scene from the graphs you received? List details such as specific patterns of peaks and where they were located.

27.7 Fire Debris Worksheet

Determine if each of the photos were arson or not. Explain and give observations.		
#1	Arson?	Explain:
#2	Arson?	Explain:
#3	Arson?	Explain:
#4	Arson?	Explain:
#5	Arson?	Explain:
#6	Arson?	Explain:

Identify the unknown accelerants using the known accelerants to compare to.	
Unknown 1	
Unknown 2	
Unknown 3	
Unknown 4	
Unknown 5	
Unknown 6	
Unknown 7	

Lab 28 Explosives

28.1 Introduction

An explosive is defined as any chemical compound, mixture, or device whose primary or common purpose is to function by explosion. Explosives can be categorized into two different types: low and high explosives. Low explosives are defined as explosives that have detonation velocities below 3,280 feet per second. Low explosives have oxygen and fuel physically mixed and must be confined in order to have an explosion. They can be ignited by a flame, spark, or shock, with the oxidation taking place relatively slowly, resulting in its main effect to be push rather than shatter objects. In contrast, high explosives express detonation rates greater than 3,280 feet per second. High explosives do not have to be confined and usually contain oxygen and fuel together in the same chemical structure. They may be sensitive to flame, spark, or shock as a primary high explosive or the explosives may need a strong impulse for being a secondary high explosive. High explosives are used with the intention to shatter objects.

A standard laboratory procedure conducted on identification of explosives is a visual examination of the substance, consisting of a microscopic and ignition susceptibility test. The microscopic examination involves using a stereomicroscope to examine the physical characteristics of the explosive.

When visually examining the explosive sample with the stereomicroscope, the physical characteristics of the explosive are noted, including, but not limited to, particle size and dimensions, size distribution, shape, color, similarity in appearance among different particles, surface roughness/coating, and porosity. Multiple particles of the explosive should be analyzed and the characteristics of the unknown explosive can then be compared to known exemplars to presumptively determine the identity of the explosive. In addition to microscopic examination, a physical examination of the questioned explosive may also include an ignition susceptibility test. The purpose of the test is to examine physical characteristics of an explosive through its reaction of combustion when exposed to an open flame. The test is conducted by exposing a single particle of the explosive to an open flame and observing the combustion reaction of the explosive. Observations of the combustion include flame color, how energetic the flame is, amount and color of smoke produced, sounds produced, other smells, color of residue, and amount of residue remaining. Like the microscopic examination, the observations of the explosive can then be compared to known explosive standards to assist in determining the identity of the unknown explosive.

28.2 Pre-Laboratory Questions

1. What are the components of an explosive?

2. What is the difference between low and high explosives?

28.3 Scenario

Last night on the other side of town a suspicious looking athletic bag was reported to be in the Wal-Mart parking lot. There was then an explosion that blew up the few surrounding cars that happened to still be in the lot. The explosion residue was brought to the crime laboratory to be examined for the possible type of bomb that was set and what type of ignition was used to set off the bomb.

28.4 Materials

- Glass watch plate
- Metal forceps
- Distilled water
- Magnify lens
- Flame source; candle or alcohol lamp

28.5 Procedure

1. Explosives analysis
   a. Examine the known explosive particles using the stereomicroscope.
   b. Record color, shape, width, and surface features of the explosive samples.
   c. Perform a burn test on an individual piece of explosive by grabbing a piece of explosive with forceps and carefully placing the particle into the flame of a lit alcohol lamp. Record details such as the flame-like color, spark/flame, smoke, and odor. *After igniting a particle, dip the forceps into distilled water and wipe them clean, ensuring the forceps are cool before testing the next particle.*
   d. Examine the unknown explosive taken from the crime scene. Repeat steps a through c for the explosive and record your data in the same way as before.
   e. Using the details you recorded about each known and unknown explosive samples, determine what type of explosive you received.

28.6 Follow-Up Questions

1. Which visual examination was more useful in presumptively determining the identity of the unknown explosive, the microscopic or ignition susceptibility test? Why?

2. What characteristics did you use in the examination of the explosives, both the microscopic and ignition test, to determine what the identity of the unknown explosive was?

28.7 Explosives Worksheet

Procedure 1a and 1b: Explosives Analysis
Provide a description about each known explosive. Include color, shape, width, and surface features.
Pyrodex
Triple Seven
Red Dot
SR 4756
Procedure 1c: Explosives Analysis
Provide a description about how each known explosive burns. Include details such as color, spark/flame, smoke, and odor.
Pyrodex
Triple Seven
Red Dot
SR 4756

(*Continued*)

Procedure 1d and 1e: Explosives Analysis
As above, provide descriptions of your unknown explosive:
Microscopic examination of unknown explosive:
Burn test of unknown explosive:

Unit 5

Forensic Physics

Lab 29

Accident Investigation

Skid to Stop Mathematics

29.1 Introduction

When a vehicle skids, the tires cease to rotate and they slide across the surface until the vehicle comes to a stop or collides with an obstacle. Accident investigators use the coefficient of friction and the average skid distance to determine the minimum speed of the vehicle at the start of the skid. The calculations in an investigation make use of the basic laws of physics, but employ different terminology and symbols for the physics variables. Calculations are made less difficult using derived equations and constants.

Accident investigation is a direct application of Sir Isaac Newton's laws of motion, work, and the law of conservation of energy. Law enforcement combines the equations used to determine kinetic energy, work done by friction, and conservation of energy into useful formulas to calculate friction and speed.

At the scene of an accident, a police officer may do multiple test skids with his vehicle. These skids are done to mimic the conditions of the accident and to use the measurements of the test skids to determine the drag factor at the site. The size of the vehicle is not a factor as mass cancels out of the equations used.

The speed of the police vehicle going into the skid to a stop is noted for each test skid. The length of each skid in feet is measured for all four wheels and then an average skid length (d) is calculated. Using the formula, $s^2 = 30df$, the drag factor (f) is determined for the each test skid and the lowest value is used. Drag factor is an indication of the amount of friction between the car tires and the road and is dependent on the type of surface. A large value for drag factor would indicate a rough surface and a low value for a smooth, slippery surface.

Once this value is determined, the officer measures the accident vehicle's skid distance in feet for all tires, takes an average skid distance (d), and then uses the formula, $s = \sqrt{30df}$, to determine the minimum speed at the start of the skid. Note that the measurements of feet and inches will need to be converted to decimals for determining the average distance (d). For example, 40 ft 4 in. = 40.33 ft.

29.2 Pre-Laboratory Questions

1. What is the drag factor?

2. Why are test skids done at an accident scene?

3. If the drag factor is low, what does that indicate?

29.3 Procedure

1. The formulas used in this exercise are as follows:
   a. Finding the drag factor (f) from test skids (d is average skid distance and s is initial speed):

$$f = \frac{s^2}{30d}$$

   b. Finding the accident initial speed (s) (d is average skid distance and f is drag factor):

$$s = \sqrt{30df}$$

2. Use the following data to determine the minimum speed of the accident vehicle when it entered a skid to stop on a level road surface. Test Skid 1 is done for you as an example.
3. Use the lowest drag factor value from the test skids in the accident speed calculation.

Test Skid 1	Measurement	Decimal (in Feet)
Right front tire	40 ft 3 in.	40.25
Left front tire	39 ft 4 in.	39.33
Right rear tire	36 ft 10 in.	36.83
Left rear tire	36 ft 5 in.	36.42

Average distance (d) = (40.25) + (39.33) + (36.83) + [(36.42)/4] = 38.21 ft.

Speedometer reading entering skid (s) = 32 mph

Drag factor (f) = $s^2/30d$ = $\frac{(32\,\text{mph})^2}{30(38.21\,\text{ft})} = 0.893$

Data	Test Skid 1	Test Skid 2	Test Skid 3	Accident Vehicle
Right front tire	40 ft 3 in.	39 ft 8 in.	43 ft 6 in.	54 ft 2 in.
Left front tire	39 ft 4 in.	40 ft 4 in.	43 ft 4 in.	54 ft 6 in.
Right rear tire	36 ft 10 in.	37 ft 6 in.	41 ft 5 in.	51 ft 10 in.
Left rear tire	36 ft 5 in.	37 ft 10 in.	41 ft 10 in.	51 ft 4 in.
Average d (ft)	38.21 ft			
Starting speed	32 mph	32 mph	34 mph	
Drag factor	0.893			f_{low} =

29.4 Work Space for Calculations

29.5 Follow-Up Questions

1. If the posted speed at the accident site was 40 mph, was the car in this exercise speeding when it skid to a stop?

2. Why would the lowest value for drag factor used in the calculation for the accident speed and not the highest or the average of the drag factors?

3. A car moves with a speed of 35 mph on a level, dry road when suddenly the driver brakes to a stop for a herd of deer crossing the road. The car stopped in 51 ft.
   a. Assuming the car and the braking force (F_f) to be constant, what would the braking distance be if the driver had doubled the speed to 70 mph?

   b. What does this tell you about safe driving habits?

   c. Calculate the coefficient of friction (drag force) for this road surface.

4. Given the following data, find the speed of the accident vehicle.

   Test skid speeds were 30 mph.

Skid test #1 data	LF	42 ft	RF	41 ft
	LR	38 ft	RR	39 ft
Skid test #2 data	LF	42 ft	RF	37 ft
	LR	40 ft	RR	39 ft
Accident skid data	LF	65 ft	RF	63 ft
	LR	59 ft	RR	62 ft
Speed ________				

Lab 30

Glass

30.1 Introduction

Rib markings, specifically on glass or hard plastic, are lines seen on the inside edges of breaks that form in the shape of a rib. These marking, also known as *Wallner lines*, are formed on the radial breaks perpendicular to the side of impact. These lines also appear as mirror images on matching edge pieces. They are formed because materials similar to glass will stretch slightly before they break. Soft plastics tend to stretch too far to make rib markings. Fractures will also develop in recognizable patterns based on where the initial impact was made. In hard objects, radial lines will form out and away from the impact point, and concentric lines will circle around the impact point, connecting the radial lines. Secondary impact points can be seen in between radial and concentric lines from a previous impact (Figure 30.1).

30.2 Pre-Laboratory Questions

1. What are radial fractures?

2. What are concentric lines?

3. What other products besides glass may also produce rib markings when broken?

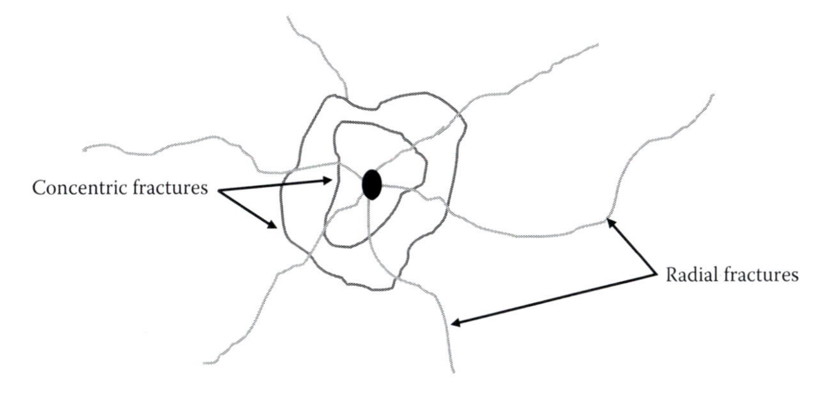

Figure 30.1
The lines of fracture due to impact on glass. The fracture lines that originate at the point of impact and move outward are *radial fractures*. The circular lines of fracture around the impact point are *concentric fractures*.

30.3 Scenario

A car was found abandoned in the school parking lot. The headlight was broken, as well as the passenger side window. Contents inside the vehicle were stolen, including the school's mascot uniform. The stolen mascot uniform was found in a student's locker. Also found in the locker were multiple pieces of glass and plastic. The analyst is asked to determine if the glass matches to the car window and the order of impact and side of impact of the broken window. The plastic found also needs to be examined to determine if it could have originated from the broken headlight.

30.4 Materials

- Magnify lens
- Broken glass
- Broken headlight

30.5 Procedure

1. Glass fracture matching
    a. Match broken glass pieces together based on their edges and surface images to complete the item. Use the clear tape to hold the pieces together.
    b. Determine if glass found fits to the broken glass found at the crime scene.
    c. Determine the order of impact and the side of impact.
    d. Sketch the final item to show the pieces as a whole.
2. Headlight
    a. Match the piece of the headlight to the headlight based on the edges of the pieces.
    b. Determine if there is a physical match between the piece and the broken headlight.
    c. Sketch the final item to show the pieces as a whole.

30.6 Follow-Up Questions

1. How do rib markings form in relation to the side of impact?

2. What do you look for when conducting glass fracture analysis?

3. If a glass bottle was sawed in half, could you make a physical match between the two pieces? Why?

30.7 Glass Worksheet

Draw out how the glass pieces were fitted back together. Label the radial and concentric fractures. Note which location was broken first.

Draw the rib marking seen on the edge of the glass. Note the side of impact.
Draw the positioning of the pieces of the headlight as the fit back together.

Lab 31

Forensic Analysis of Glass

Identifying the Type of Glass

31.1 Introduction

In the world of crime scene investigation, glass is considered as trace evidence. Trace evidence is one that can be linked to a crime scene. It can be found at the scene or on someone or something that had been at the scene. Locard's exchange principle states that when two objects come into contact, there is a mutual exchange of matter between the objects. That matter is trace evidence. Broken glass can travel large distances and if the fragments are small, they can easily embed into hair, clothing, shoes, or other objects.

Glass breakage can occur during various types of crimes. Burglary, breaking and entering, assaults, and hit-and-run car-pedestrian accidents are common crimes where glass evidence can furnish important clues in the investigation. Therefore, investigators need to be able to distinguish between tempered, window, and household glass.

When collecting window glass samples from a scene, it is important to take samples from the intact source of the glass for comparison purposes. The investigators must note on the samples as to which surface represents the outside of the glass and which is the inside. Any glass fragments not intact (scattered about a room) should be collected. Investigators will look at those fragments for dirt on one side, putty or paint in order to determine the orientation of the glass before it was broken. In this activity, we will learn how to look at fracture patterns to determine the direction of force on these samples. As a result of this study, the investigator can determine whether the trauma occurred inside or outside the home.

Besides determining the type of glass at a scene, the investigator studies the fracture patterns in the broken glass. When more than one fracture is present, it is possible to determine the sequence that the breaks occurred. Together with other types of evidence from a crime scene (such as fingerprints, toolmark and footprint impressions, bloodstain patterns, DNA evidence, and chemical analysis) the forensic investigator pieces together the puzzle to re-create a logical sequence of events supported by evidence. All these clues are important steps leading to the reconstruction of events at a crime scene.

In this series of glass activities, you will learn how the intrinsic properties of glass (density and index of refraction) can be used to separate common types of glass. The density of a material depends upon the mass and volume of a substance. The formula for density is

$$\text{Density} = \frac{\text{Mass}}{\text{Volume}}$$

The density of a very small piece of glass can be determined using an electronic balance to measure the mass in grams and the volume in milliliters. Volume is found by suspending the small glass fragment in a beaker of water and finding the *mass* of water displaced by the glass. This then translates to the *volume* of water displaced by the glass, because water has a density of 1 g/ml and therefore the mass and volume are equal. The volume of water displaced by the glass equates to the volume of the glass, since it was the substance that displaced the water.

When light travels from one medium to another of different density, it does not travel in a straight line, but bends at the boundary between the two mediums. This bending of light is called *refraction*. Light refracts because the density difference changes the velocity of the light. Light slows down when it travels into a high density material, and the light ray bends toward the *normal* line. (A normal line is a line draw perpendicular to the boundary between the two media) See Figure 31.1, Diagram A. Conversely, light speeds up when it travels from a high density to a low density material, and the light ray bends away from the *normal* line. See Figure 31.1, Diagram B.

Index of refraction (n) is a ratio that compares the speed of light in a vacuum to the speed of light in the material:

$$n = \frac{\text{Velocity of light in a vacuum}}{\text{Velocity of light in the material}}$$

The index of refraction of air is approximately 1.0003. Since liquids and solids are denser than air, light will travel slower in them. This will make the index of refraction greater in liquids and solids.

Light rays can be used to find the index of refraction of an unknown substance. By projecting a laser light through a liquid and then out into the air, the amount of refraction can be measured. Angles are measured and then the index of refraction is calculated using Snell's law:

$$(n_1)(\sin\theta_1) = (n_2)(\sin\theta_2)$$

where:

n_1 is the index of refraction of first medium
n_2 is the index of refraction of the second medium
θ_1 is the *angle of incidence* between the incoming ray and the normal line
θ_2 is the *angle of refraction* between the outgoing ray and the normal line

See Figure 31.1, Diagram C.

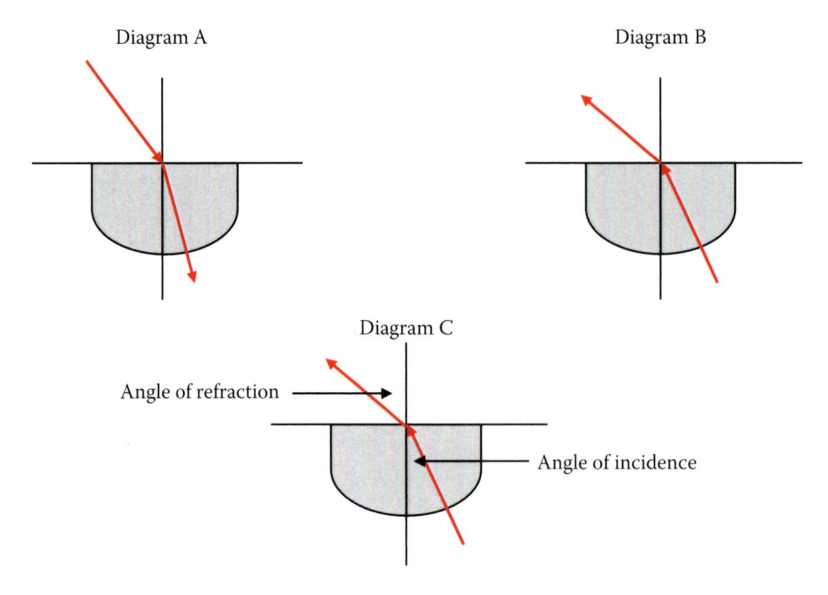

Figure 31.1
Diagrams A–C showing the refraction of light as it travels through materials of different density.

31.2 Pre-Laboratory Questions

1. What are the two physical properties of glass that will help in the determination of the type of glass found at a crime scene?

2. What happens to light when it travels from one medium to another of different density?

3. Why is the analysis of glass important in crime scene investigation?

31.3 Scenario

A robbery occurred at a pharmaceutical research facility. The suspect had a weapon and used it to break into the building. A window of the lab was broken and also some of the laboratory equipment. There were glass pieces all over the counters and floors of the drug research lab. One suspect has been taken into custody. The suspect had trace evidence of glass on his shoes and clothing. Samples were collected and will be analyzed in this activity.

31.4 Materials

Part One Materials

- Broken glass samples: Leaded glass, tempered glass, window glass, Pyrex glass
- Crime scene samples of glass
- 50 ml beaker
- Water
- Sewing thread
- Scissors
- Forceps
- Electronic scale with sensitivity to 0.01 or 0.001 g
- Safety goggles and gloves

Part Two Materials

- Penlight laser or other linear light source
- White unlined 8½ × 11 in. paper
- Transparent half-circle dish
- Sharpened pencil

- Ruler
- Protractor
- Water
- Vegetable oil
- Clove oil
- Cinnamon oil
- Safety goggles and gloves

Part Three Materials

- 4–60 mm Petri dishes or 50 ml beaker
- Small, flat, known glass fragments (tempered, lead crystal, Pyrex, window glass)
- Crime scene samples
- Refractive liquids: water, vegetable oil, clove oil, and cinnamon oil
- String or forceps
- Hand lens
- White paper
- Safety goggles and gloves

31.5 Forensic Analysis of Glass Laboratory Exercises

31.5.1 Activity One: Determining the Density of a Piece of Glass

1. Obtain one sample of each of the four glass types. Using the forceps, pick up the glass, place it on the electronic scale, and measure the mass. Be very precise in your measurements and procedure. Record the masses in a data table (Table 31.1).
2. Carefully tie a piece of thread, 10 cm long, around each glass sample. Fill the 50 ml beaker three-fourths full of water. Place the beaker and water on the scale and re-zero or tare it, so that it now reads 0. While the beaker is still on the balance, carefully lower one of the pieces of glass into the water so that the entire piece of glass in is the water *but not touching the sides or the bottom of the beaker.* It should be suspended in the middle of the water. Record the reading as the mass of the water displaced in the data table. Repeat this procedure for each of the glass samples.

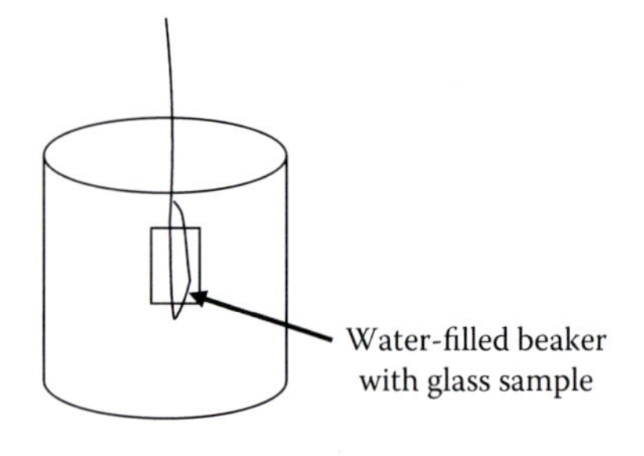

3. The mass of the water displaced by the glass can be used to calculate the volume of the glass.
4. The density of water is 1.000 g/ml. Therefore, the numerical value for the mass of water and the numerical value for the volume of water would be the same number. Record the volume of water displaced in the data table.
5. The volume of water displaced is the same as the volume of the glass, record this volume in the data table as the volume of the glass.
6. Calculate the density of each glass fragment and record in the data table.

$$\text{Density} = \frac{\text{Mass}}{\text{Volume}}$$

7. Repeat the same procedure for the crime scene glass samples to determine density and compare to the known samples (exemplars).

31.5.2 Activity Two: Determining the Refractive Index of Liquids

1. Draw vertical and horizontal lines with the pencil on the white paper as shown below.
2. Mark the point where the two lines cross with a darkened dot. Draw a line that is 30° from the lower vertical line (normal line) and intersects the dot as shown in the diagram. This will be the incident light ray.
3. Line up the half-circle container as shown below, so that the midpoint of its straight edge lines up with the darkened dot.

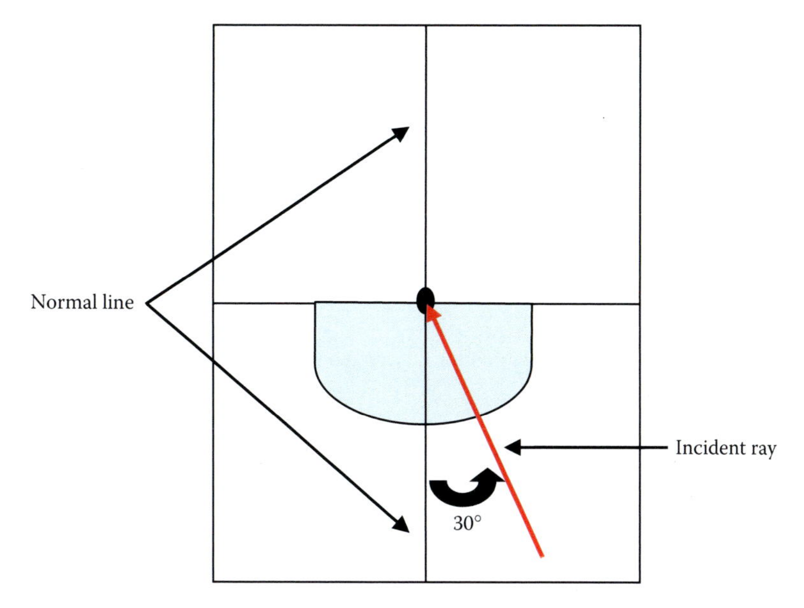

4. Half fill the container with water. With the room darkened, line the laser pen along with the incident ray line, so that it shines through the water and intersects with the dot.

   Safety note: Do not shine the laser light directly into the eyes.
5. Mark the position of the laser light ray that exits the water and travels through the air (the refracted light ray). Use three dots to mark its position.

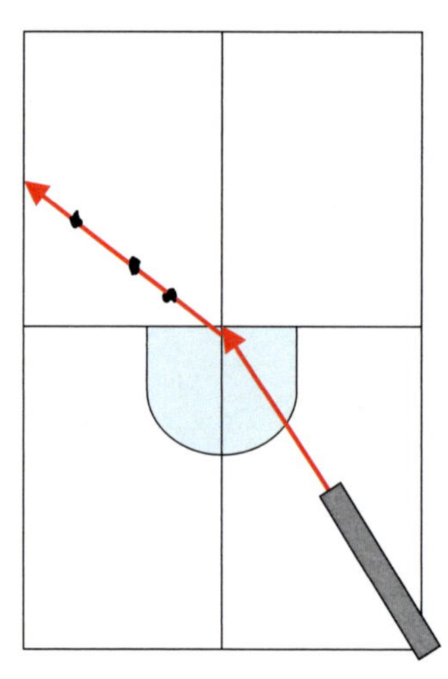

6. Remove the laser. Draw the refracted ray by connecting the three dots and the center dot with a straight line.
7. Measure the angle between the vertical line (normal line) and the refracted ray.
8. Record the angles for both rays in Table 31.2. Label the paper *Water Refractive Index*, and set aside. Clean and dry the container.
9. Repeat steps 1 through 8 for the remaining liquids.
10. Use the following Snell's law to find the index of refraction of the liquid.

$$(n_1)(\sin\theta_1) = (n_2)(\sin\theta_2)$$

11. Optional calculation: Calculate the percent difference between experimental refractive index and known refractive indices for the liquids tested in the activity.

31.5.3 Activity Three: Identification of Glass Fragments Using Index of Refraction

1. Place the four Petri dishes or beakers onto the white paper. Write the names of the liquids under the dishes as shown below.

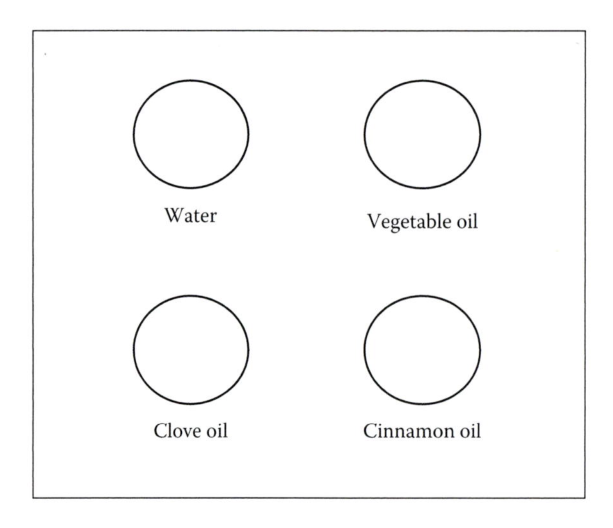

2. Carefully pour a small amount of the liquid into the containers.
3. Using forceps, place the tempered glass into the center of the container. If the glass is not covered by the liquid, then carefully add more liquid into the dish until the glass sample is completely covered.
4. Note in Table 31.3 if the sample is *visible, somewhat visible,* or *not visible.* Use the hand lens if necessary.
5. Carefully remove the glass with the forceps and dry the sample. Repeat the process for the remaining liquids. Note results in Table 31.3.
6. Repeat steps 3 through 5 for the other three glass samples. These will be the exemplars.
7. Using the known refractive indices of the liquids shown below, and the summary Table 31.4, determine the possible refractive index of each sample, or perhaps what the refractive index is not.

Known Refractive Indices	
Water	1.333
Vegetable oil	1.474
Clove oil	1.52–1.54
Cinnamon oil	1.60–1.62

8. Test the crime scene samples and compare to the exemplars to determine the type of glass (Table 31.4).

31.6 Follow-Up Questions

1. What type of glass was found at the crime scene? Give evidence to support your conclusion.
2. Is glass evidence class or individual evidence? Why?
3. Think of an instance where glass can be specifically tied to a crime scene and describe.
4. List three ways glass evidence is important trace evidence in crime scene investigation.

31.7 Forensic Analysis of Glass Worksheets

Glass Fragment	Mass of Glass (g)	Mass of Water Displaced (g)	Volume of Water Displaced (ml)	Volume of Glass (ml)	Density of Glass (g/ml)
Window					
Pyrex					
Tempered					
Lead crystal					
Crime scene A					
Crime scene B					
Crime scene C					

Determining Index of Refraction of a Liquid

Liquid Being Tested	Angle of Incidence (θ_1)	Angle of Refraction (θ_2)	Index of Refraction of Liquid (n_1)	Index of Refraction of Air (n_2)
Water	30°			1.0003
Vegetable oil	30°			1.0003
Clove oil	30°			1.0003
Cinnamon oil	30°			1.0003

Immersion Results for Known Glass Samples (Visible, Somewhat Visible, Not Visible)

Glass Fragment	Water	Vegetable Oil	Clove Oil	Cinnamon Oil
Tempered				
Lead crystal				
Pyrex				
Window glass				

Summary Table for Refractive Index of Glass and Crime Scene Samples

Glass Sample	Disappears Mostly In	Visible In	Possible Refractive Index
Tempered			
Lead crystal			
Pyrex			
Window glass			
Crime scene A			
Crime scene B			
Crime scene C			

Lab

32

Measuring the Diameter of a Human Hair

32.1 Introduction

When water waves pass through a small opening or barrier, they spread out or diffract. The diffracted waves can interfere with each other, producing a new wave pattern due to the waves adding or canceling each other. In a similar manner, light, which has wave properties, diffracts after passing through an opening. This produces a pattern of light and dark bands on a screen placed in the beam's path. In this activity, you will use laser light and diffract the light around a hair placed in its path. The pattern produced on the screen will look like the one as shown in Figure 32.1, with a bright central spot (central maximum) and surrounding bands of light and with spaces between that have no light (dark bands).

To calculate the approximate width of a human hair, measurements should be taken of the distance from the hair to the screen (D), the distance between the centers of the first dark bands in the diffraction pattern (y), and the average wavelength of the laser (λ). See Figure 32.2 for location of the first dark bands. The number one (1) in the formula is indicative that these are the first dark bands measured from the center of the diffraction pattern (the central maximum). The width of the hair (a) is calculated using the following formula. *Note that all measurements should be in meters.*

$$a = \frac{(1)(\lambda)(D)}{(y/2)}$$

$$\text{Width of the hair} = \frac{(1)(\text{wavelength})(\text{hair-to-screen distance})}{(\text{distance between centers of dark bands}/2)}$$

32.2 Pre-Laboratory Questions

1. What causes the laser light to make the light and dark band patterns?

2. Why must all the measurements be converted to meters?

3. How are human hairs different from each other?

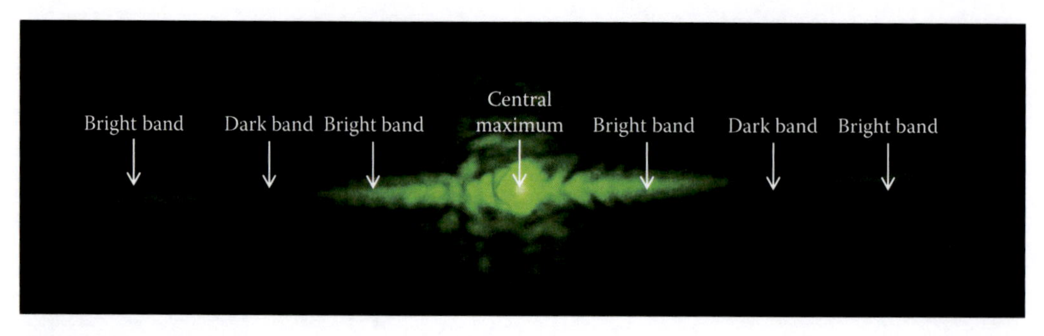

Figure 32.1
Diffraction pattern caused by shining a green laser through a human hair.

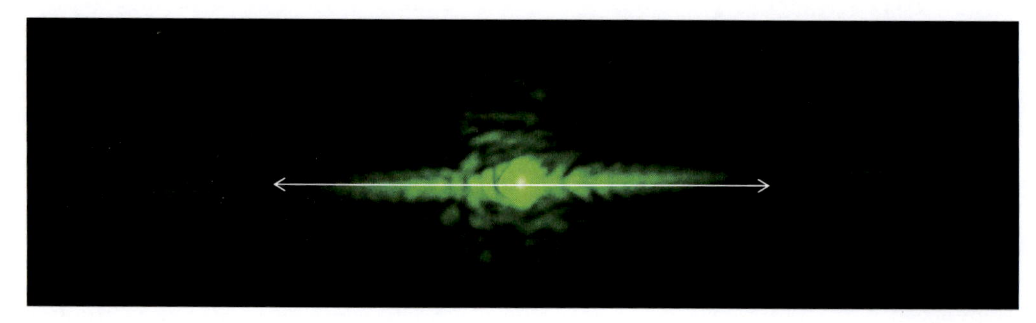

Figure 32.2
Measuring the distance between the centers of the first dark bands.

32.3 Scenario

Upon investigating an assault of an elderly man, crime scene investigators discovered a suspicious hair on the man's jacket. The hair did not seem to match the victim, who had white hair. The hair will be analyzed using diffraction to determine its diameter, which will aid in identification of the type of hair.

32.4 Materials

- Laser
- Metric tape measure
- 6- or 12–in. ruler with metric rule
- Calculator
- Screen (poster board, chalkboard, or whiteboard)
- Slide holder with opening
- Tape

32.5 Procedure

Safety note: Laser light should never enter the eye. Never look into a laser beam and do not point the laser at anyone else's eyes. Be cautious on highly reflective surfaces, as they may reflect the laser beam into the eye.

1. Secure a sample of head hair.
2. Mount the hair across the middle of the opening of the slide holder, pulling the hair taut and straight. Tape the hair securely into place.

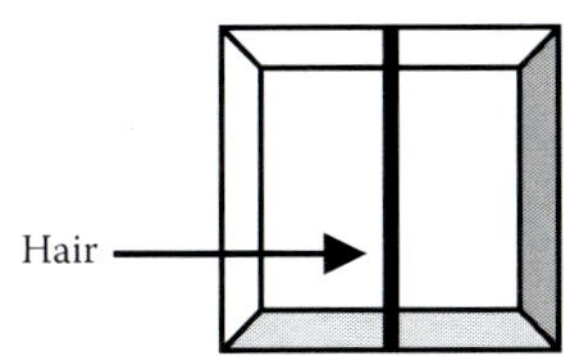

3. Place the screen 5 or more meters away from the hair slide.
4. Secure the laser either by mounting it on a stand or placing it on a stack of books.
5. Hold the slide without moving it in front of the laser. This might be done using another stack of books or using a ring stand.
6. The room needs to be darkened in order to see the pattern.
7. Shine the laser light through the opening, so that it shines on the hair and a pattern is captured on the screen. The distance from the slide to the screen may have to be increased in order to clearly see the light and dark patterns. Figure 32.3 shows the proper placement of lab materials for this investigation.
8. Carefully mark the position of the bright bands on the screen. Remove the laser and measure the distance between the centers of the first dark bands (y) on either side of the bright central maximum. Record in Table 32.1. Be sure the measurement is converted to meters.
9. The wavelength of the laser should be printed on the side of the laser. If the laser wavelength is given as a range, find the average wavelength in meters and record in the table (1 m = 1,000,000,000 nm).
10. Measure the distance from the hair to the screen in meters (D).
11. Use the formula to find the approximate width of the human hair (a). Note that the distance y must be divided by two in the formula.
12. Convert the diameter of the hair in meters to micrometers (1 m = 1,000,000 μm).
13. Repeat the procedure for each hair in your group.

Figure 32.3
Apparatus setup

32.6 Follow-Up Questions

1. Why does the formula for finding the diameter of the hair state that it is the approximate diameter?

2. Compare hair diameter with other students. Are the diameters similar or different? State any reasons for the similarities and/or differences.

3. Using the literature values for the accepted diameter of your type of hair, calculate the percent difference between your experimental diameter and the accepted literature value.
Discuss your results.

4. Try a different human hair—from an arm, facial hair, or an eyelash. How do they compare to the head hair?

32.7 Measuring the Diameter of a Human Hair Worksheet

Diameter of a Human Hair

Source of Hair	Wavelength (λ)	Hair Screen (D)	Band to Band (y)	Hair Diameter (a)

Bibliography

Bodziak, W., *Footwear Impression Evidence, Detection, Recovery, and Examination*, 2nd edition, CRC Press, Boca Raton, FL, 2000.

Houck, M. and Siegel, J., *Fundamentals of Forensic Science*, 2nd edition, Academic Press, Boston, MA, 2010.

Jones, P. and Williams, R., *Crime Scene Processing and Laboratory Workbook*, CRC Press, Boca Raton, FL, 2009.

Keppel, R., Brown, K., and Welch, K., *Forensic Pattern Recognition: From Fingerprints to Toolmarks*, Pearson, Upper Saddle River, NJ, 2007.

Kirk, P.L., *Crime Investigation: Physical Evidence and the Police Laboratory*, Interscience Publishers, Inc., New York, 1953.

Kubic, T. and Petraco, N., *Forensic Science Laboratory Manual and Workbook*, 3rd edition, CRC Press, Boca Raton, FL, 2009.

Lentini, J., *Scientific Protocols for Fire Investigation*, CRC Press, Boca Raton, FL, 2006.

Saferstein, R., *Basic Laboratory Exercises for Forensic Science*, Pearson, Upper Saddle River, NJ, 2007.

Siegel, J. and Mirakovits, K., *Forensic Science: The Basics*, 2nd edition, CRC Press, Boca Raton, FL, 2010.

Wheeler, B. and Wilson, L., *Practical Forensic Microscopy: A Laboratory Manual*, Wiley-Blackwell, Chichester, 2008.

Index

Note: Locators followed by "*f*" and "*t*" denote figure and table in the text

Q

R

S

T

U

V

W